A Review of the Air Force Fitness Assessment

SEAN ROBSON, ISABEL LEAMON, MARIA C. LYTELL, MIRIAM MATTHEWS, MARGARET CHAMBERLIN

Prepared for the Department of the Air Force
Approved for public release; distribution unlimited

RAND PROJECT AIR FORCE

For more information on this publication, visit www.rand.org/t/RRA762-1

Library of Congress Cataloging-in-Publication Data is available for this publication.
ISBN: 978-1-9774-0669-9

Published by the RAND Corporation, Santa Monica, Calif.

Cover: LSOphoto/Getty Images/iStockphoto.

Preface

Military readiness requires service members to be mentally and physically fit to perform mission- and job-related duties in a wide variety of environments. To determine whether U.S. Air Force (AF) personnel are maintaining physical fitness, a fitness assessment (FA) is administered to all airmen. The AF's Force Management Policy Directorate (AF/A1P) requested an evaluation of the relevance and comprehensiveness of FA components to ensure readiness, support the National Defense Strategy, and promote a culture of health and well-being across the AF. This report provides an overview of research relevant to the FA components, identifies potential gaps, and offers recommendations for improvement. This report describes work that should be of interest to military policymakers and researchers involved in setting and evaluating military physical fitness standards.

The research reported here was commissioned by the Air Force's Force Management Policy Directorate (AF/A1P) and conducted within the Manpower, Personnel, and Training Program of RAND Project AIR FORCE as part of a fiscal year 2018–2019 project *Review of Evidence Relevant to the Air Force Fitness Assessment*.

RAND Project AIR FORCE

RAND Project AIR FORCE (PAF), a division of the RAND Corporation, is the Department of the Air Force's (DAF's) federally funded research and development center for studies and analyses, supporting both the United States Air Force and the United States Space Force. PAF provides DAF with independent analyses of policy alternatives affecting the development, employment, combat readiness, and support of current and future air, space, and cyber forces. Research is conducted in four programs: Strategy and Doctrine; Force Modernization and Employment; Manpower, Personnel, and Training; and Resource Management. The research reported here was prepared under contract FA7014-16-D-1000.

Additional information about PAF is available on our website:

www.rand.org/paf/

This report documents work originally shared with DAF on February 5, 2020. The draft report, issued on February 14, 2020, was reviewed by formal peer reviewers and DAF subject-matter experts.

Contents

Figures

Tables

Summary

Issue

To ensure that service members are sufficiently physically fit to serve, the U.S. Air Force (AF) has established a variety of medical and physical standards. As part of its fitness program, the AF administers a fitness assessment (FA) to all airmen that contains four components: (1) a 1.5-mile run or 2.0-kilometer walk, (2) abdominal circumference (AC) measurement, (3) push-ups, and (4) sit-ups. RAND Project AIR FORCE was asked to evaluate the relevance of these tests to ensure mission readiness and support the National Defense Strategy.

Approach

Our study team conducted its evaluation of the AF-FA using scientific evidence drawn from published literature on relevant fitness components, with an emphasis on the potential for current assessments to meet overall health and deployment requirements. Evidence from the literature was augmented with workshops and discussions with a variety of subject-matter experts, including those familiar with deployment readiness training.

Findings

Our results suggest that the AF-FA has several strengths and a few potential gaps that could be addressed with dedicated resources.

- Overall, the current AF-FA is a practical assessment that measures critical components of health-related fitness using well-supported assessments.
  - The 1.5-mile run is a valid and well-supported measure of cardiorespiratory fitness. Alternative tests, such as the shuttle run, bike test, and row ergometer, are also valid measures and may have utility in specific cases.
  - AC (or waist circumference) is a valid measure of body composition. Alternative measures, such as the waist-to-height ratio, may provide some additional benefits beyond AC.
  - Sit-ups and push-ups are acceptable measures of muscular endurance. However, there are concerns about subjectivity associated with evaluating these tests and the risk of injury associated with sit-ups.
  - Muscular strength is not currently measured in the AF-FA, but including it should be considered to ensure that airmen can perform common military tasks during deployment.

- Flexibility is not measured by the AF-FA, but it is not clearly linked to health outcomes, injuries, or military task performance.
- The AF does not fully address the physical fitness of airmen for advanced deployments, specifically to hostile or uncertain environments.

Recommendations

To address potential gaps in the current AF-FA, we offer the following recommendations:

- *Conduct a trial study to explore alternative assessments*. Many alternative tests could be considered. In addition to determining the reliability and validity of alternative assessments, the cost-benefit of changes needs to be considered.
- *Leverage AF data to establish criterion-referenced standards based on health risks (or other important outcomes) for all fitness components*. Linking relevant data would allow the AF to establish meaningful cutoff scores directly tied to health risk and readiness.
- *Consider developing a predeployment FA*. Many movement patterns and fieldcraft training tasks might require physical abilities not covered by AF-FAs (see Figure S.1). Reports from training instructors suggest that up to 30 percent of airmen arrive at training without the requisite fitness. A new assessment would require the exploration of such topics as the standard setting for an FA, who should take the test and how often, and what tests should be included.

Figure S.1. Subject-Matter Expert Ratings of Movement Patterns and Fieldcraft Training Tasks Not Sufficiently Covered by the Air Force Fitness Assessment

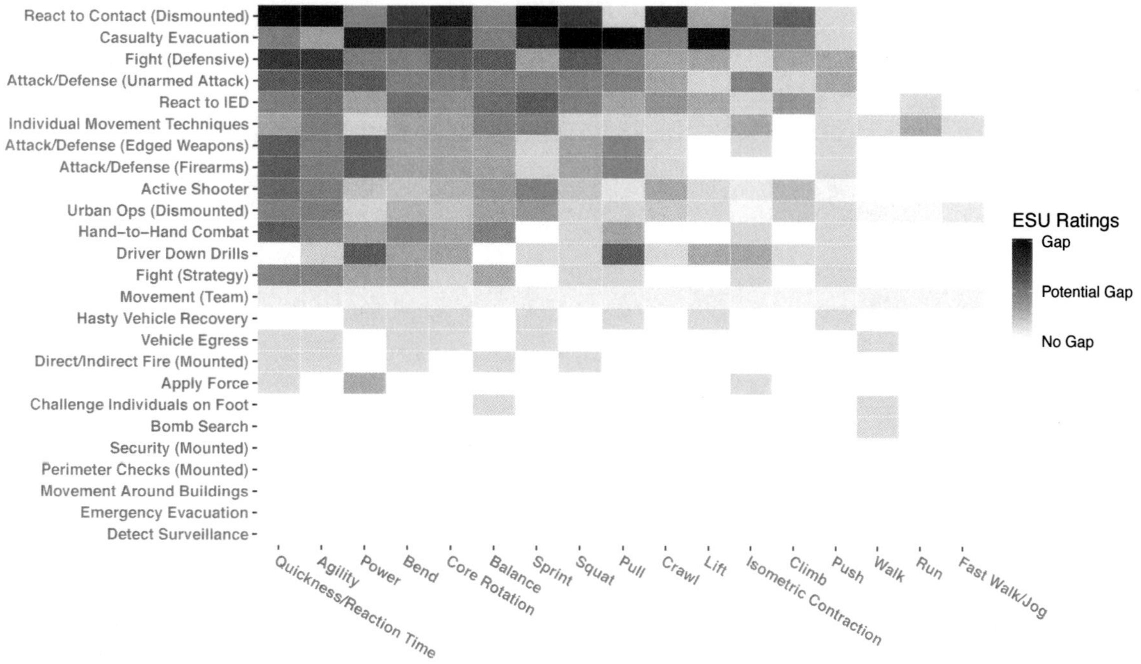

NOTE: ESU = Exercise Science Unit; IED = improvised explosive device; ops = operations.

Acknowledgments

Several people supported this project. We begin by thanking our sponsors, Brigadier General Robert LaBrutta and Brigadier General Troy Dunn. We also express gratitude to our action officer, Emi Izawa, who assisted our efforts. We also thank Denise Harris, SMSgt. Larry Anderson, TSgt. Justin Wadekamper, and Ken Phillips for their assistance in coordinating discussions and providing documentation. We also appreciate the fieldcraft course instructors, Neal Baumgartner, and members of the Exercise Science Unit for providing ratings and feedback on the deployment training tasks. Baumgartner was also instrumental in providing a location to host the first subject-matter expert workshop.

We are very appreciative of the time and inputs provided by all of the subject-matter experts who attended the workshops. We are also thankful to the fieldcraft course instructors and the 421st Combat Training Squadron for supporting our site visit to observe training and conduct discussions with trainees and instructors.

We are grateful for our RAND Corporation colleagues, including Paul Emslie, who provided data on advanced deployment training; Ray Conley, former director of the RAND Project AIR FORCE Manpower, Personnel, and Training Program, who provided guidance and support throughout the project; and our peer reviewers. Finally, we thank our reviewers Paul Mayberry and Patricia Deuster for their thoughtful comments and feedback to improve the quality of this report.

Abbreviations

AC	abdominal circumference
ACSM	American College of Sports Medicine
AF	Air Force
AF-FA	Air Force Fitness Assessment
AFFMS	Air Force Fitness Management System
AFI	Air Force Instruction
AFSC	Air Force specialty code
BMI	body mass index
CDC	Centers for Disease Control and Prevention
CRE	cardiorespiratory endurance
DoD	U.S. Department of Defense
ESU	Exercise Science Unit
FA	fitness assessment
FC	fieldcraft
FC-CR	Fieldcraft Contingency Response
FC-H	Fieldcraft Hostile
FC-U	Fieldcraft Uncertain
FFI	Fit-Fat Index
FMS	Functional Movement Screen
OPAT	Occupational Physical Ability Test
SLJ	standing long jump
SME	subject-matter expert
TCCC	Tactical Combat Casualty Care
WC	waist circumference
WHR	waist-to-hip ratio
WtHR	waist-to-height ratio

1. Introduction

Military readiness encompasses service members' ability to perform mission- and job-related duties in a wide variety of environments and requires service members to be mentally and physically fit (Constable and Palmer, 2000). To ensure that service members have the requisite physical fitness to serve, the U.S. Air Force (AF) and its sister services have established various medical and physical standards. These standards are first applied as part of an initial screening for military entrance (e.g., at a Military Entrance Processing Station), and then, after joining, service members must continue to maintain fitness in accordance with their respective service policies.

Definitions of Physical Fitness

There is wide consensus that there are at least two main categories of physical fitness: (1) health-related fitness and (2) performance-based (or skill-based) fitness (Caspersen, Powell, and Christenson, 1985). This differentiation is also referred to as *Tier 1 fitness* and *Tier 2 fitness* by the AF Exercise Science Unit (ESU) and other military organizations (Palmer et al., 2000; Robson et al., 2017). Although we focus on Tier 1 fitness, we briefly introduce the concept of Tier 2 fitness, and its relationship to Tier 1 fitness, in this chapter.

Tier 1 Fitness

The primary objective of Tier 1 fitness is to promote general health, which parallels scientific guidelines that promote physical activity as an important strategy (2018 Physical Activity Guidelines Advisory Committee, 2018). Embracing health as a primary objective in its fitness program, the AF strives to "motivate all members to participate in a year-round physical conditioning program that emphasizes total fitness, to include proper aerobic conditioning, muscular fitness training, and healthy eating" (Air Force Instruction [AFI] 36-2905, 2015, p. 7).

Although the AF does not explicitly list all health conditions targeted by the fitness program, previous research has demonstrated that physical activity and fitness have been linked to a wide variety of diseases and chronic health conditions, including weight status (e.g., obesity),[1] all-cause mortality, cardiometabolic conditions (e.g., heart disease, stroke, hypertension, type 2 diabetes), certain types of cancer (e.g., bladder, breast, colon, kidney, stomach, lung), and brain health (e.g., dementia, depression) (2018 Physical Activity Guidelines Advisory Committee, 2018).

In addition to health as a primary objective, the AF fitness program is designed to increase productivity and decrease absenteeism. To support these goals, commanders and supervisors are expected to promote a culture of fitness and create "an environment for members to maintain physical fitness and health to meet expeditionary mission requirements" (AFI 36-2905, 2015, p. 7). In summary, the AF Tier 1 fitness program broadly supports military readiness by minimizing potential chronic health conditions, which subsequently reduces health care costs and improves productivity by reducing absenteeism.

Tier 1 fitness has several characteristics that differentiate it from Tier 2 fitness (see Table 1.1). First, Tier 1 applies to all airmen and may require different standards for men and women and different age groups to achieve the same level of general health. For example, 1.5-mile run times would need to be different for men and women to achieve the same level of health risk for cardiovascular disease. Similarly, the Institute of Medicine concluded in its review that, "on average, women have a higher percent body fat than men. Weight gain and lifestyle changes during the childbearing and childrearing years, as well as the hormonal and metabolic changes that accompany pregnancy and menopause, are associated with higher body fat. Thus, the gender-specific fat standards are appropriate" (Institute of Medicine, 2003, p. 11).

---

[1] There is a long history in developing and refining categories to differentiate healthy and unhealthy weight levels. The current concepts and definitions used to classify people as "overweight" or "obese" are based on body mass index (BMI), which is calculated using a person's height and weight. The Centers for Disease Control and Prevention (CDC) lists four classifications based on BMI (see CDC, 2020):

- Underweight = <18.5
- Normal weight = 18.5–24.9
- Overweight = 25–29.9
- Obesity = BMI of 30 or greater.

Table 1.1. Differences Between Tier 1 and Tier 2 Fitness

Fitness Tier	Primary Objectives	Secondary Objectives	Target Population	Tests	Standard
Tier 1	• Reduce health risks • Promote culture of fitness	• Increase productivity • Decrease absenteeism	• All airmen	• 1.5-mile run • Abdominal circumference (AC) • Push-ups • Sit-ups	• Based on projected health outcomes (e.g., health or injury risk) • May vary by age and gender
Tier 2	• Ensure capability to perform job-specific physically demanding tasks	• Reduce risk of job-related injuries	• Physically demanding AF specialty codes (AFSCs)	• Varies based on job demands	• Same standard for those performing same job duties • Based on projected job performance outcomes

SOURCE: Adapted from Robson et al., 2017.

Second, Tier 1 fitness does not include skill-based components, such as agility or reaction time, which may be needed to perform job-specific tasks (e.g., maneuver under fire).[2] Although there may be some variation in how health-related fitness (Tier 1) is defined,[3] there are generally five primary components: cardiorespiratory endurance (CRE), muscular endurance, muscular strength, flexibility, and body composition (see Table 1.2) (American College of Sports Medicine [ACSM], 2017; Institute of Medicine, 2012).

---

[2] The Army's Occupational Physical Ability Test (OPAT) and Army Combat Fitness Test are two examples that are consistent with the Tier 2 fitness construct. These assessments are designed to determine a soldier's capability to perform tasks specific to different Army Military Occupational Specialties.

[3] For example, the Institute of Medicine (2012) also considered balance as a health-related fitness component.

Table 1.2. Health-Related Fitness Components

Fitness Component	Definition
CRE	The ability of the circulatory and respiratory systems to supply oxygen during sustained physical activity.
Body composition	The relative amount or percentage of different types of body tissue (i.e., bone, fat, muscle) that are related to health.
Muscular endurance	The ability of a muscle group to execute repeated contractions over a period of time sufficient to cause muscular fatigue or to maintain a specific percentage of the maximum voluntary contractions for a prolonged period of time.
Muscular strength	The ability of a muscle group to develop maximal contractile force against a resistance in a single contraction.
Flexibility	The ability to move a joint through its complete range of movement.

SOURCES: ACSM, 2017; Caspersen, Christenson, and Pollard, 1986.

Air Force Fitness Assessment (Tier 1)

According to internal AF documents, the AF fitness assessment (FA) has been revised several times over the past four decades. Between 1981 and 1992, the AF measured CRE using a 1.5-mile run. In 1992, the AF switched to a cycle ergometry test known as the bike test to address concerns about the safety of the 1.5-mile run. The AF reinstated the 1.5-mile run in 2004 in response to the AF Chief of Staff's desire to adopt an FA that promotes a warrior culture. At this time, the AF also introduced its first muscular endurance tests (push-ups and sit-ups).

Between 1985 and 2004, body composition was primarily measured using a combination of height and weight. In 2004, the AF began using AC as its primary indicator of body composition. The most recent modification to the AF-FA was made in 2013 when the AF replaced the 1-mile walk with the 2.0-kilometer walk as the alternative cardiorespiratory assessment for airmen receiving a medical exemption from the 1.5-mile run.

The AF currently administers an FA to all airmen twice a year, with the exception of airmen scoring an "Excellent" who only need to take the FA once a year. The FA has four components: (1) a 1.5-mile run or 2.0-kilometer walk, (2) AC measurement, (3) push-ups, and (4) sit-ups. The run or walk is designed to measure cardiorespiratory fitness, the AC measurement is a body composition measure, and push-ups and sit-ups assess muscle fitness.

Tier 2 Fitness

Tier 2 fitness tests and standards are designed to ensure that individuals can perform physically demanding tasks and duties required by their occupational specialty. Tier 2 standards are different from Tier 1 in several ways, including how test standards are applied (Table 1.1). Generally, Tier 2 standards are the same for all individuals in a specialty to ensure that everyone has the capability to perform critical job tasks. The distinction between Tier 1 and Tier 2 fitness is described in recent policy guidance for Tactical Air Control Party (TACP) and Air Liaison Officers (ALO):

> All US Air Force Airmen must maintain a necessary level of physical fitness to meet the science based standards of the Air Force Fitness Assessment. This assessment, referred to as a *Tier 1* physical fitness test, is designed with health criterion standards to ensure Airmen are present for duty in good health and general fitness. However, a Tier 1 level of fitness does not necessarily reflect specific military task achievement. Some Air Force (AF) occupations or AF Specialty Codes, including ALO and TACP, require higher and broader levels of physical fitness to meet the physical demands of their operational mission sets. Thus, such specialties need a set of physical fitness tests and standards based on their AFSC-specific physical duty requirements, we refer to these as Tier 2 occupationally- specific, operationally-relevant (OSOR) tests and standards (Department of the Air Force, 2018, p. 2).

Predeployment Fitness

One area not explicitly covered by either Tier 1 or Tier 2 is physical fitness for deployment. Currently, the AF does not have a specific predeployment FA. However, depending on the types of tasks performed and the environment (e.g., hostile location), airmen may need different fitness levels to perform their deployed mission. Although Tier 1 assessments may, in some cases, be relevant for measuring the capability to meet deployed mission demands (e.g., running), some deployments may require additional physical abilities (e.g., agility to move in response to enemy contact). Noting this potential gap, the ESU suggested that the term *Tier 1-D*[4] could be used to refer to a potential predeployment FA. For the remainder of this report, we use the term *Tier 1-D* to refer to predeployment fitness required of all airmen.

Our Approach

Several sources informed our evaluation of the AF-FA and its relationship to general health and fitness for deployment. We briefly describe these sources in this chapter and provide additional details on methodology and sources in subsequent chapters and in Appendixes A through C. Our evaluation was focused on two primary objectives of the AF-FA: (1) determine how the AF-FA relates to general health, and (2) determine whether the AF-FA can be extended to measure the physical fitness needed for deployment missions. The AF Tier 1 program is also designed to promote a culture of fitness. However, this specific objective is not systematically evaluated as part of this effort but will be explored as part of a separate research effort. We also

---

[4] Although the purpose of a predeployment FA may fit closer to a Tier 2 construct, the AF has used *Tier 1-D* to generally refer to a predeployment FA that all airmen may need to complete. The AF could consider relabeling to *Tier 2-D* if the assessment is only relevant for a subset of airmen (e.g., those deploying to certain environments).

do not focus on the specific AF standards. However, we note that the standards for a health-based assessment may need to vary based on age and gender to ensure that the same level of health or injury risk is achieved. Using different standards is not contrary to U.S. Department of Defense (DoD) guidance, which requires gender-neutral occupational standards (Hardison, Hosek, and Bird, 2018). That is, standards for a Tier 2 assessment must be gender-neutral (i.e., a single standard for men and women). Similarly, a predeployment FA designed to ensure that airmen can perform physically demanding tasks should also be gender-neutral.

Research Literature Review

This report is grounded in scientific evidence drawn from published literature on the fitness components relevant to the AF's Tier 1 FA, with an emphasis on the potential for current assessments to serve as indicators of general health. For the purpose of this report, we define *general health* as the level of risk for developing one or more chronic health conditions or diseases and risk of sustaining an injury. In addition to general health, we explore research on the relationship between Tier 1 assessments and general military task performances to inform the analysis of the need to develop a Tier 1-D FA. For each fitness component, we address the following questions:

1. What evidence is available to support the inclusion of each fitness component into an AF-wide FA?
2. What is the relationship between each fitness component and health, general military task performance, and injury risk?
3. What assessments are appropriate for measuring each fitness component?
4. What limitations or concerns are associated with the FAs?
5. Are there alternative tests that should be considered?

To address the first two questions, we focus primarily on studies with evidence of criterion-related validity. That is, we review studies that evaluate how well the fitness components and assessments predict health, injury, and performance outcomes. We broaden our criteria to address questions 3–5 to include implementation issues, such as administration cost, practicality for mass testing, equipment requirements, and acceptability (e.g., perceived fairness).

We occasionally incorporate research on Tier 2 task or job-specific FAs to the extent that they provide additional evidence for the inclusion of one or more FAs, although these factors are not explicitly included in our searches. The review provides an overview of the research literature and is not designed to be comprehensive. When available, we highlight prior quantitative systematic reviews (i.e., meta-analyses) and other systematic reviews. In particular, studies in our review were prioritized using the following criteria:

- Included at least one of the five fitness components.
- Provided data on the relationship between fitness and relevant outcomes, including injuries and health symptoms, diseases, and risks.

- Provided a quantitative review of primary studies (e.g., meta-analyses).
- Provided a systematic qualitative review of primary studies.
- Used a military-specific population in primary study.
- Tracked adults over time in longitudinal primary study.

Target Populations

The research literature on health-related fitness is broad and often focuses on the general population rather than military-specific populations. It is important to interpret research in light of any potential differences that might exist between airmen and the general population. For example, we present recommended minimum standards when supported by research, but the extent to which these standards are meaningful to meet AF objectives remains to be evaluated in future research using data on the health of airmen. This type of evaluation is especially important because all airmen are prescreened prior to joining the AF and must meet certain physical medical requirements. Therefore, on average, airmen are likely to be healthier and more fit compared with the general population. Despite these potential differences, the broader literature is appropriate to determine relationships between FAs and risks of developing specific health conditions or diseases.

Although we focus heavily on research using adult populations, we reference some research that has evaluated fitness measures for the youth population (Mahar et al., 2014; Morrow et al., 2009; Pate and Daniels, 2013; Stodden, Sacko, and Nesbitt, 2017). This research base is sometimes more comprehensive because of the broad interests among educational institutions and society to promote physical fitness for all children and adolescents. Therefore, we consider these studies to the extent that findings are potentially informative in considering available fitness tests or could guide future research within the AF.

Workshops with Subject-Matter Experts

In addition to the research literature, we held workshops and discussions with various subject-matter experts (SMEs). Two workshops were conducted with a total of 17 SMEs representing a variety of backgrounds and perspectives (e.g., policy, safety, medical) from the AF, DoD, and private industry. These workshops were designed to collect systematic feedback about the current AF-FA, strengths, limitations, and alternative assessments to address any potential gaps. Additional detail on the workshops is provided in Appendix A.

Deployment Readiness Training and Subject-Matter Expert Ratings

To inform our evaluation of predeployment fitness, we held discussions with SMEs familiar with deployment readiness and training (e.g., course developers, instructors, and leadership). We also observed some parts of training that required physical effort and conducted brief focus groups with students following completion of the training exercises. Finally, we conducted a preliminary evaluation of the physical demands required during advanced deployment training.

More specifically, this evaluation involved collecting ratings of training tasks and movement patterns (e.g., push, pull, bend) completed by nine course instructors and seven personnel from the ESU. Additional details on the methodology for this analysis are provided in Chapter 3.

Criteria for Evaluation

Using the information sources described earlier, we considered several criteria (Table 1.3) to inform our evaluation of the AF-FA and recommendations (Chapter 4). Whenever possible, we prioritized research provided in systematic reviews to inform validity evaluations. Other criteria (e.g., practicality, acceptability) required inputs from SMEs to more fully understand assessment implementation and administration within the context of resources available to the AF.

Table 1.3. Criteria and Sources Used in Our Evaluation of the Air Force Fit Assessment

		Information Source			
Criteria	**Definition[a]**	**Systematic Review**	**Primary Research**	**SMEs**	**SME Ratings**
Criterion-related validity evidence	The degree to which an assessment (e.g., 1.5-mile run) is statistically related to outcomes (e.g., diabetes diagnosis)	X	X		
Convergent validity evidence	The degree to which relationships between assessment scores (1.5-mile run) and other measures of the same or related construct (e.g., maximal oxygen uptake or VO_2 max) are positive		X		
Comprehensiveness	The degree to which assessments measure important physical abilities (e.g., CRE)			X	X
Reliability	The degree to which scores are consistent over one or more potential sources of error (e.g., time, raters, conditions of measurement) in the application of a measurement procedure	X	X	X	
Practicality	The costs and ease of administration and implementation of assessments			X	

		Information Source			
Criteria	Definition[a]	Systematic Review	Primary Research	SMEs	SME Ratings
Acceptability	The degree to which test-takers, test administrators, and policymakers believe in and can support the FAs			X	
Risk of injury	The degree to which preparing for and completing an FA contributes to or exacerbates an injury		X	X	

[a] Definitions for validity and reliability were adapted from Society for Industrial and Organizational Psychology, 2018.

Additional Criteria Not Considered

The criteria presented in Table 1.3 are not exhaustive, and the AF will need to consider other important criteria by analyzing its own existing data or by collecting new data.[5] Three particularly relevant criteria include potential test bias, evidence supporting standards and scoring systems, and utility to the AF. Although some of these criteria can be informed by prior research, interpretations will be sensitive to differences in how the full assessment is designed and implemented. Therefore, we briefly introduce these criteria next and recommend that the AF more thoroughly consider them in future research.

Potential test bias is a complex issue and requires evaluating statistical relationships between assessments and outcomes across different subgroups (e.g., gender, age, race). Predictive bias occurs when one group's (e.g., women) scores on an outcome measure are systematically under- or overpredicted. In other words, the assessment does not predict outcomes equally well for men and women. This could create an unfair system when one single test standard or cutoff is used for both subgroups because the standard would have a different interpretation for men compared with women. Such predictive bias is not necessarily present when absolute differences are observed in assessment measures by subgroups (e.g., women scoring lower on pull-ups does not necessarily imply predictive bias). Predictive bias analyses could help the AF evaluate its scoring system.

Other important issues to consider are determining the optimal number and combination of tests, when to use different stratifications (e.g., age groups), and how the scoring system shapes airmen's behavior to maintain or improve their fitness. Finally, the utility of the AF-FA can be affected by length of service, among other factors. The AF-FA will likely have more utility (e.g., potential to reduce disease and related health conditions) when retention levels are high and

[5] For additional criteria to evaluate a Tier 2 assessment, see Robson et al., 2018.

airmen have more years of total service. Therefore, the AF may need to evaluate the effectiveness of the AF-FA across different years of service and other potentially relevant factors (e.g., occupational specialty).

Organization of This Report

The remaining chapters in this report summarize the research and workshop SME perspectives on health-based physical fitness (Chapter 2), evaluate the relevance of the AF-FA to fitness for deployment (Chapter 3), and conclude with recommendations and possible courses of action for improving the AF-FA to promote overall health and readiness of all airmen (Chapter 4).

2. Health-Based Fitness Assessments, Outcomes, and Limitations

This chapter integrates the research review and workshop SME perspectives on each fitness component and a variety of assessments that can be used to measure each component. Specifically, we define each fitness component, provide an overview of assessments, evidence for inclusion in an AF-wide FA, and potential limitations with current assessments. To determine the evidence for inclusion, we consider three general questions. First, we evaluate whether the fitness component is related to general health (i.e., physical conditions or diseases). Second, we evaluate how the fitness component is related to musculoskeletal injuries. Finally, we evaluate whether the fitness component is related to general military task performance. This latter form of evidence is used to explore whether the AF-FA components can be used to inform decisions about an airmen's fitness for deployment. Therefore, the two primary outcomes influencing our evaluation of the Tier 1 AF-FA are general health and injuries.

We begin this chapter by providing an overall summary of the research findings in Table 2.1. The remainder of this chapter discusses these findings and considerations for specific assessments in more detail. We conclude with a section summarizing additional SME perspectives on issues related to physical fitness in the AF.

Table 2.1. Summary of Research Findings

Fitness Component	Primary Findings	Example Assessments	Considerations
CE	• Predicts many health outcomes, including cardiovascular disease, all-cause mortality, type 2 diabetes, and cancer mortality • Predicts performance on a variety of general military tasks • Poor performance on timed run tests are associated with higher rates of musculoskeletal injury	1.5-mile run	• Safety and space requirements
		Shuttle run	• May underestimate VO_2 max
		Cycle ergometry test	• Less face valid • Administration complexity
Body composition	• Predicts a broad variety of health outcomes, including hypertension, type 2 diabetes, dyslipidemia, and metabolic syndrome • Much less research on relationships between body composition, general military task performance, and injuries	Waist circumference (WC)	• Concerns about fairness of waist measurement • Not sensitive to stature • Different cutoffs for men and women
		Waist-to-height ratio (WtHR)	• Concerns about fairness of waist measurement • Single recommended cutoff value that applies to both men and women
		Waist-to-hip ratio (WHR)	• Concerns about fairness of waist measurement • Not sensitive to stature
		BMI	• May misclassify certain types of individuals
Muscular fitness	• Comparatively less research linking muscular fitness to health outcomes compared with CRE and body composition • Muscular fitness is critical to safe and effective performance of physically demanding task • More research is needed to evaluate effectiveness of different core endurance assessments (e.g., plank)	Upper body • Push-ups, handgrip strength, and pull-ups Core • Sit-ups Lower body • Sprints, standing long jump (SLJ), and vertical jump	• Validity of assessments varies depending on the general military task being performed
Flexibility	• No consistent relationships noted with general health, musculoskeletal injuries, or general military task performance	Not recommended	• Limited to no predictive benefit for this fitness component

Cardiorespiratory Endurance

As described in Chapter 1, CRE is one of the five health-related physical fitness components recognized by the ACSM (ACSM, 2013; ACSM, 2017). CRE[6] reflects "the ability of the circulatory and respiratory systems to supply oxygen during sustained physical activity" (ACSM, 2017). Laboratory assessments are available to assess maximal oxygen consumption, but more-practical field expedient tests have also been developed and validated that can effectively measure CRE.

Types of Assessments

From 1992 to 2003, the AF measured CRE using the cycle ergometer (or bike test). In 2004, the AF replaced the cycle ergometry with the 1.5-mile run to address the Chief Secretary of the AF's concerns that the cycle ergometry was "not consistent with thc growing demands of our warrior culture" (Callander, 2004). Another major change occurred in 2010 when the AF updated its 1.5-mile run standards using published data from the Cooper Center Longitudinal Study of The Cooper Institute (Blair et al., 1989; Farrell, 2018). These data were used as a basis to establish different gender- and age-specific cut scores that are associated with different levels of health risk. More specifically, different cut scores are used for men and women and across ten-year age groups (younger than 30, 30–39, 40–49, 50–59, and 60-plus years of age) to determine low, moderate, and high health risk for each airman.

Evidence for Inclusion

As described next, the research literature provides strong evidence for the inclusion of specific CRE assessments on the basis of general health benefits, prediction of performance in general military tasks, and reducing injury risks. In particular, the 1.5-mile run, an important component of the current AF-FA, is a well-supported measure of CRE and provides the foundation for performing physically demanding military tasks. Furthermore, workshop SMEs noted that the run is simple and familiar and that equipment is not a barrier.

General Health Benefits

CRE has received extensive support as an important predictor for many health outcomes, including cardiovascular disease, all-cause mortality, type 2 diabetes, and cancer mortality (Kodama et al., 2009; Barry et al., 2014; Zaccardi et al., 2015; Schmid and Leitzmann, 2015). Research further suggests that CRE can mitigate some of the risks associated with higher levels

---

[6] There are several related terms, such as *cardiorespiratory fitness*, *aerobic fitness*, and *cardiovascular endurance*, which have different technical definitions and measurements. However, we considered these terms collectively in our review and use the term *cardiorespiratory endurance* (or *CRE*) as a broad label to refer to the general body of research.

of body fat (Stevens et al., 2002) and is a stronger indicator of mortality compared with physical activity (Myers et al., 2004). Barry et al. (2014) concluded in their meta-analysis comparing fit (high CRE) with unfit (low CRE) individuals that "[c]ompared to normal weight-fit individuals, unfit individuals had twice the risk of mortality regardless of BMI" (p. 382).

Performance of General Military Tasks

Many common military tasks have physical demands that require higher levels of CRE, which is a strong predictor of performance in various physically demanding tasks and is more strongly linked to general military task performance than other fitness components are (Friedl et al., 2015). In a recent quantitative review, Hauschild and colleagues reported that cardiorespiratory tests demonstrated strong to very strong correlations with a variety of military tasks, including digging, climbing, moving fast, carrying a stretcher, repeated lifting and lowering, lifting and carrying, and crawling (Hauschild et al., 2014). Another recent study on the performance of common military tasks, conducted for the Canadian military, demonstrated that CRE, measured by the 20-meter shuttle run, accounted for a significant portion[7] of the variance in total task performance (Tingelstad et al., 2016). The relative importance of high CRE can be attributed to individuals' ability to utilize a lower percentage of their maximal aerobic capacity, which results in lower physiological stress, reduced fatigue, and reduced injury risks (Knapik et al., 2001). Finally, research has shown that individuals below a certain VO_2 max were less likely to complete a 30-minute walk while wearing body armor (Ricciardi, Deuster, and Talbot, 2007).

Injury Risk

A recent systematic review of CRE and musculoskeletal injuries found evidence supporting a relationship between injury risk and timed runs over a set distance, such that poorer performance on these runs was associated with higher musculoskeletal injury risk (Lisman et al., 2017). Other types of tests (such as the shuttle run, a test in which individuals run back and forth between two lines as fast as possible) were found to predict injury risk in men but were inconclusive for women because of insufficient research. And several other tests (e.g., timed step test, progressive endurance run) were found to have limited or mixed support. However, evidence from the Army's recent implementation of the OPAT suggests that men and women who performed fewer shuttles[8] during the test sustained a greater injury rate during the first ten weeks of Initial Entry Training (Hauret et al., 2018).

---

[7] Partial R^2 values for the shuttle run ranged from .32 (male) to .36 (female).

[8] Shuttles were counted as part of the Interval Aerobic Run (i.e., beep test), which is an incremental, multistage shuttle run that requires individuals to run between two lines 20 meters apart.

Evaluation of Specific Measures

Timed distance runs, such as the 1.5-mile and the 2.0-mile runs, are strongly associated with maximal oxygen uptake, or VO_2 max, and can be implemented with minimal equipment (i.e., track, stopwatch). Another strong measure of CRE is the shuttle run, as recently implemented in the Army's OPAT. Shuttle run tests require less space than longer-distance runs (Wilkinson et al., 2014). Although research generally suggests that shuttle run tests are an effective measure of CRE, some researchers have suggested that they underestimate VO_2 max (Cooper et al., 2005). Despite these concerns, a recent meta-analysis established a moderate to strong relationship between the shuttle run and VO_2 max and further suggested that it is a reasonable estimate of CRE (Mayorga-Vega, Aguilar-Soto, and Viciana, 2015). Consequently, a shuttle run test may be considered as an acceptable alternative when space constraints prevent the longer time-based runs.

Other CRE tests, such as the bike test (cycle ergometer), the rower test (row ergometer), and the 2.0-kilometer walk, could be considered in more-limited cases (e.g., injury) (Jorgensen et al., 2009; Hauschild et al., 2014). Currently, airmen who receive a medical exemption[9] from the 1.5-mile run may be eligible to take the alternative 2.0-kilometer walk. Although the walk is not as strong an indicator of CRE, research suggests that it is still a valid assessment (Weiglein et al., 2011). To improve the validity of the walk assessment, Vickers (2002) suggests that several other factors, including gender, weight, age, and heart rate, must be used when estimating VO_2 max. Other alternative tests, such as the bike and rower tests, require equipment, which may present additional challenges. Because of the costs of purchasing and maintaining the equipment for these tests, implementation AF-wide might not be practical. Furthermore, these tests may also lack face validity in that airmen may not immediately see a link between these tests and military tasks and requirements. Despite these concerns, workshop SMEs suggested several benefits to the row ergometer as an alternative for the 1.5-mile run. These benefits included easy calibration (within a few minutes to calibrate up to eight machines), lower impact on the lower body, high objectivity, and relatively simple to administer. A few other concerns were raised, including a need for clear guidelines for medical practitioners to determine who is safe to perform the row, uncertainty on the potential benefits (i.e., how many airmen could do the row but not run?), and the need for equipment that would limit how many could be tested at one time.

Potential Limitations

Increased physical training that includes running can increase the risk of injury (Knapik et al., 2006). There appears to be a slight paradox in that very little prior physical activity or very

---

[9] The alternative cardiorespiratory assessment is considered when airmen have "either musculoskeletal or clinical (e.g., cardiac, pulmonary, etc.) conditions that preclude running," but are otherwise eligible to walk (AFI 36-2905, 2015, p. 61).

high running mileage can both increase the risk of developing a musculoskeletal injury (Kaufman, Brodine, and Shaffer, 2000). Consequently, a balance in the amount of physical training is needed to minimize the risks for developing an injury. Unfortunately, as Nindl (2012) concluded, "[t]here is a paucity of research that has considered physical performance adaptations over the 'life cycle spectrum' of the Warfighter, particularly among operational units. It is difficult to determine the optimal physical training programs to implement without additional validation studies" (p. 7). One approach to minimizing potential risks associated with running is to limit the total distance over a set time period and to consider placing individuals into different ability groups based on their fitness levels (Knapik et al., 2006).

Workshop SMEs also identified a few limitations of the 1.5-mile run, including concerns with injuries because (1) training is not consistent (e.g., running too far too fast, cramming before the test) and (2) running is a high-impact activity. These injuries were cited as one cause for high rates of exemption from the run test. To address injury concerns, SMEs suggested that proper education and training were needed.

There were also concerns that some airmen have difficulty with pacing, but one SME suggested that providing feedback on lap times can help improve pacing (e.g., maintaining a similar pace for each lap). At least two SMEs suggested the need for a cold-weather alternative, such as the beep test. Finally, there was some discussion that the 1.5-mile run benefits runners and that there are other ways, such as cycling, to develop and maintain cardiorespiratory fitness.

Summary

CRE is a critical component of any FA designed to promote general health and performance on physically demanding military tasks. There are several assessments that could be considered to measure CRE; however, the 1.5-mile test used by the AF is an excellent measure that is easy to administer and well supported by research. When the 1.5-mile run cannot be used because of injury or poor weather conditions, there are several alternative assessments that could be considered, including the shuttle run, 2.0-kilometer walk, and bike test. Although the rower test may have several advantages, more research is needed about potential implementation challenges (e.g., medical guidelines for establishing eligibility of injured airmen to take the rower test in place of the 1.5-mile run).

Body Composition

Body composition addresses relative amounts of different types of body tissue, including bone, fat, and muscle (ACSM, 2017). As discussed in this chapter, the importance of body composition for measuring health-based fitness has been widely accepted in the academic and professional literature. Despite the volume of research, there is no single best measure of body composition. Furthermore, some research suggests that there may be very little difference in how well body composition measures predict cardiovascular disease risk (Emerging Risk Factors Collaboration, 2011).

Types of Assessments

Although laboratory tests (e.g., hydrodensitometry) are available to measure body fat, we focus on measures that have been or could be easily implemented by the military services (Wagner and Heyward, 1999). Common measures include BMI, WC, WtHR, and WHR (Friedl, 2012; Griffith et al., 2018). The AF-FA measures WC using AC.[10] The maximum allowable AC for men is 39 inches and 35.5 inches for women.

Evidence for Inclusion

The existing literature is conclusive that there are clear linkages between body composition and general health. There is far less research, however, examining the relationships between body composition, military task performance, and injuries. Furthermore, this research suggests that these relationships are more complex. For example, one research study found that the strongest women who have higher WC with lower health risk (e.g., lower WC) may perform less well on physically demanding military tasks (Friedl, 2012). However, less body fat is generally associated with better performance on other physical fitness tests, which correlate with better task performance (Crawford et al., 2011). The relationship between BMI and injuries also appears to be more complex. For example, Knapik (2015) summarizes previous research that found that injury risks are higher for those with either low BMI or high BMI. That is, the relationship between BMI and injuries may be nonlinear.

General Health Benefits

Previous research showed that, relative to normal BMI, obesity was generally associated with significantly higher all-cause mortality. The results from this study suggest that increased risks of mortality are restricted to those with the highest BMI levels (Flegal et al., 2013). A more recent systematic review suggests that there are increased risks of heart failure incidence and mortality in both the overweight and obese categories (Aune et al., 2016). Another meta-analysis evaluating the relationships between overweight, obesity, and 18 co-morbidities extracted from 89 prospective studies demonstrated increased risks for certain types of cancer (e.g., colorectal, kidney), cardiovascular diseases (e.g., hypertension, coronary artery disease, stroke), and other health conditions (e.g., asthma, chronic back pain) (Guh et al., 2009).

---

[10] AFI 36-2905 defines *AC* as "[a] circumferential measure of abdominal girth at the iliac crest that is positively and highly correlated with internal fat and in turn disease risk independent of body mass" (AFI 36-2905, 2015, p. 60).

Evaluation of Specific Measures

We evaluate four specific measures of body composition. The first three: 1) WC, (2) WtHR, and (3) WHR each require measuring WC. Consequently, these assessments are somewhat more susceptible to measurement error compared with BMI, the fourth measure we considered. Although BMI can be a more objective assessment, using only height and weight, it can misclassify certain individuals as being obese or overweight. We also discuss a fifth measure, the Fit-Fat Index (FFI), which is a composite measure combining WC and CRE. We discuss additional details about these measures in the following sections.

Waist Circumference

A large WC is an indicator of excessive abdominal fat, which increases health risks. In a study that included measures of both BMI and WC, individuals with higher WC had higher likelihood of hypertension, type 2 diabetes, dyslipidemia, and metabolic syndrome across all BMI categories: normal weight, overweight, and class I obese (Janssen, Katzmarzyk, and Ross, 2002). Furthermore, the associations between WC and health outcomes often remained significant after adjusting for the confounding variables (e.g., age, race, poverty-income ratio, physical activity, smoking, and alcohol intake). This evidence suggests that WC, or AC, is an important indicator of health risk. In the following sections, we review several alternatives to WC, including BMI, WHR, WtHR, and the FFI.

Waist-to-Height Ratio and Waist-to-Hip Ratio

The relatively simple calculation of WtHR is conducted by dividing WC by height. The boundary values for WtHR are the same for both genders as and for all ethnic groups (Browning, Hsieh, and Ashwell, 2010). Across genders and ethnic groups, a single WtHR cutoff of over 0.5 is critical, and signifies an increased risk of negative health outcomes (Browning, Hsieh, and Ashwell, 2010). The WHR is similarly straightforward, conducted by dividing WC by hip circumference (Bener et al., 2013). However, WHR cutoffs are not the same for men and women. The World Health Organization recommended that cutoff points indicating substantially increased risk of metabolic disease are above 0.9 for men and 0.85 for women (World Health Organization, 2011).

Research consistently shows that the WtHR and WHR are more predictive of metabolic conditions, diabetes, and cardiovascular disease than the BMI (Browning, Hsieh, and Ashwell, 2010; Dalton et al., 2003). Research also suggests that WtHR may be superior to WC at estimating the risk for future poor health outcomes. One systematic review combining results from 31 studies found that WtHR was a better measure than either WC or BMI for predicting cardiometabolic risk factors for both men and women (Ashwell, Gunn, and Gibson, 2011).

Both WtHR and WHR are easy to implement, requiring no additional equipment over the WC measure. Although research generally finds stronger support for WtHR and WHR compared with WC, research is not entirely conclusive. Some studies found WC to be more effective as a measure of fitness, though these were in the minority.

Body Mass Index

BMI is a crude measure of body composition and has some important limitations. First, BMI cutoffs used to diagnose obesity have high specificity, but low sensitivity to identify adiposity. That is, cutoffs may fail to identify half of the people with excess body fat percentage (Okorodudu et al., 2010). The CDC notes that "at an individual level, BMI can be used as a screening tool but is not diagnostic of the body fatness or the health of an individual" (CDC, 2020). Rothman (2008) further notes that "as a measure of body fat . . . BMI has serious flaws" (p. 556). More specifically, BMI does not differentiate between fat mass and other characteristics contributing to a person's weight, such as muscle and bone. Consequently, muscular individuals can be misclassified as being overweight or even obese. Furthermore, at any given BMI level, the actual percentages of body fat vary considerably. Therefore, airmen with widely different body fat percentages may have the exact same BMI. Recognizing some of these limitations, Meadows et al. (2018) suggests that using BMI to estimate military rates of overweight and obesity may be imprecise, especially for military personnel who are physically fit. Meadows further notes the possibility that "muscular service members may have been misclassified as overweight or obese" (p. 47). Consequently, any conclusions based solely on BMI may be insufficient to guide policy decisions about any specific airman's health and fitness.

Despite these limitations, BMI has been associated with many health outcomes. For example, high levels of BMI are associated with increased risks of cancer (Moghaddam, Woodward, and Huxley, 2007), all-cause mortality (Aune et al., 2016; Global BMI Mortality Collaboration et al., 2016; Flegal et al., 2013), and coronary heart disease (Mongraw-Chaffin et al., 2015). However, as presented earlier, other assessments, such as WtHR, generally appear to be stronger indicators of disease and related health conditions.

Fit-Fat Index

Prior research has acknowledged that the relationship between body composition and fitness can be complex and some individuals with higher body fat may still perform very well on all other fitness components (Leu and Friedl, 2002). The FFI, a more recently developed measure, was designed to more accurately capture these intersections between body composition and physical fitness (Sloan et al., 2018). Because of its novelty, we focus our discussion of this measure on the methodology and its potential utility to the AF.

The FFI was proposed in 2016 with the goal of combining a measure of CRE with WtHR to create a more nuanced and accurate measure of fitness relative to body composition. For the FFI, CRE is expressed as estimated maximal metabolic equivalents, based on total duration of a symptom-limited maximal modified Balke graded exercise test (the patient walks on a treadmill to exhaustion, at a constant walking speed, while gradient/slope is increased every one or two minutes). The FFI is then calculated by dividing maximal metabolic equivalents by WtHR. Scores commonly range from 10 to 50 on a continuous scale, with higher scores being better (Sloan et al., 2016).

The literature around FFI is limited but growing. Analysis of a U.S. national sample found that increased FFI was associated with reduced risk of all-cause mortality (Frith and Loprinzi, 2017a; Frith and Loprinzi, 2017b). In a study of men in the U.S. Navy, higher levels of FFI were independently and more consistently associated with having average or better health-related quality of life (physical and mental) than other known predictors were (Sloan et al., 2015).

The potential benefit of the FFI stems from the use of combined measures of fitness and central adiposity to produce a composite measure that is more effective than measures of adiposity alone. This concept is one that the AF could adapt to create its own measurement or index.

Potential Limitations

These body composition measures have several potential limitations, which primarily concern validity, reliability, and fairness. First, BMI does not discriminate between muscle mass and fat and potentially misclassifies individuals as overweight or obese. WtHR and WHR are potentially more predictive of disease risks compared with WC. However, more research is needed to directly compare these assessments across diseases and within military populations. Some assessments may be perceived to be less fair. The WC requires using different standards (cutoffs) to achieve similar risk levels for men and women and does not take into account a person's height.

Research on the FFI is promising, but more research is needed to determine whether a composite of CRE and WC can provide value beyond measuring these components separately. Another concern is that the FFI requires a treadmill, which could be impractical for large-scale testing. Therefore, the AF would need to explore creating a new composite by combining 1.5-mile run times with WC.

Furthermore, as noted by workshop SMEs, the added complexity could make the FFI more difficult to explain to service members. This greater complexity might also increase the risk that the FFI would not be fully accepted by airmen. More evidence on the validity of the FFI in military populations, and the feasibility of implementing such a measure, is needed before it is proposed for implementation.

Summary

Body composition is an important component of a health-based FA. There are several effective measures that predict health risk, each with their own advantages and disadvantages. BMI is objective and the simplest measure to implement using only height and weight but may misclassify airmen as being overweight or obese. WC and WHR may be better indicators of health risks but can be perceived as unfair and do not take into account a person's stature. WtHR has similar strengths and limitations to WC and WHR but does incorporate height to address stature. Research also suggests that no single measure would be best in all cases—therefore, some combination of body composition measures might be needed to maximize prediction of health risk. Along these same lines, the FFI is a composite measure combining WC with CRE that requires further research to address measurement and implementation challenges.

Muscular Fitness

Muscular fitness is typically divided into two categories: muscular endurance and muscular strength. As discussed in Chapter 1, muscular endurance encompasses the ability to sustain or repeat a physically strenuous motion over a short period of time, and muscular strength is the ability to exert a certain amount of force in a single physically strenuous motion (ACSM, 2017; Palmer and Soest, 1997). In this chapter, we review the relevance and validity of muscular fitness measures. Our review found some evidence that general muscular fitness provides overall health benefits and more evidence that muscular fitness of specific body regions may be important for performance of general military tasks and injury prevention; however, not all assessments are equally reliable or practical to implement. Another important area that remains to be addressed is establishing a meaningful standard on any muscular fitness test that is associated with reduced health risks. We discuss the point further in Chapter 4.

Types of Assessments

It is difficult to assess full-body muscular fitness with just one measure, so tests of muscular fitness typically focus on one region of the body: upper body, lower body, or core/trunk (Hauschild et al., 2014). Laboratory assessments of muscular endurance and strength often require free-weights or weight machines (e.g., leg press, incremental dynamic lift, bench press). This review, however, focuses on more field expedient measures, such as push-ups, sit-ups, jump tests, and a few other common field tests. Within the AF-FA, muscular fitness is currently assessed with the push-up and sit-up tests.

Evidence for Inclusion

The existing literature provides evidence for the inclusion of certain muscular FAs on the basis of general health benefits, prediction of performance in general military tasks, and possible injury prevention.

General Health Benefits

There is some research linking muscular fitness with general health benefits. A review of the literature found that musculoskeletal fitness is correlated with bone health and bone density (Pate, Oria, and Pillsbury, 2012; Palmer and Soest, 1997). Similarly, participation in resistance training programs can lead to increases in bone mass (Palmer and Soest, 1997). These outcomes are primarily important for preventing health complications, such as osteoporosis in middle- and older-aged populations (Pate, Oria, and Pillsbury, 2012; Palmer and Soest, 1997).

Muscular fitness may reduce metabolic risk for populations of all ages. Greater muscle mass correlates with higher basal metabolic rate (Palmer and Soest, 1997). Skeletal muscle has been found to help regulate glucose and fat metabolism, and skeletal muscle fitness may be associated with risk factors for developing metabolic syndromes (Pate, Oria, and Pillsbury, 2012). Muscular strength, specifically, may also be associated with disease outcomes and mortality: a large, international longitudinal study found that reduced muscle strength, as measured by handgrip strength, is associated with the increased risks of cardiovascular disease and all-cause mortality (Leong et al., 2015). Although most of these studies are correlational, they suggest that muscular fitness may be relevant for promoting positive health outcomes. One recent exception is a longitudinal study covering a ten-year period among male firefighters, which found that baseline push-up capacity predicted future incidence of cardiovascular disease (Yang et al., 2019).

Performance of Military Tasks

Muscular fitness is also important for performance of military tasks. There have been few studies on the importance of general or total-body muscular fitness in completing military activities, so these analyses predominantly focus on the muscular fitness of specific body regions. Hauschild and colleagues conducted a systematic review and meta-analysis of 26 studies that evaluated correlations between physical fitness test performance and ability to execute 12 relevant military tasks (i.e., repeated lift and lower, single lift and lower, lift and carry, stretcher carry, casualty drag, push/pull, loaded march, move fast, climb, crawl, dig, and multiactivity) (Hauschild et al., 2014). The list of military tasks was developed based on a review of the physical duties associated with military service in various North Atlantic Treaty Organization countries and performance in certain physically demanding jobs (i.e., police officers and firefighters).

They found that 83 percent of military tasks (ten out of 12) had at least moderate correlations ($r \geq 0.40$) with upper-body endurance, 78 percent with lower-body endurance (seven out of nine), 67 percent with lower-body and upper-body strength (eight out of 12), 63 percent with core strength (five out of eight), and 18 percent with core endurance (two out of 11; Hauschild et al., 2014). This meta-analysis suggests that there is limited evidence to include core endurance in physical fitness tests on the basis of assessing military task performance. Many of the military tasks did not have sufficient data on their relationship to core strength, so additional research may be necessary before generating a conclusion regarding the applicability of core strength to performance of military job tasks. Strength and endurance of the upper and lower body are all relevant for completing physically demanding military activities.

Injury Risk

Research on the importance of muscular fitness for injury prevention is somewhat less conclusive and also varies by body region. A systematic review of 45 studies examining physical fitness and injury found that push-up performance was negatively associated with injury risk for men but not for women (de la Motte et al., 2017). In 15 out of 22 studies, men's push-up performance negatively correlated with injury in multi- or univariate models. For women, however, only seven out of 16 studies indicated a univariate relationship between push-ups and injury (de la Motte et al., 2017). Overall, this review suggested that better push-up performance is associated with reduced injury risk for men. But for women, the evidence is more limited (de la Motte et al., 2017; Nye et al., 2016; Jones et al., 1993; Roy et al., 2014). Insufficient research has addressed the correlation between pull-ups and injury (de la Motte et al., 2017).

Findings regarding the relationship between core endurance and injury are mixed; there are fewer studies demonstrating significant relationships. De la Motte and colleagues report a multi- or univariate relationship between sit-ups and injury in men for five out of 24 studies (de la Motte et al., 2017). For women, four out of 16 studies in the de la Motte review (2017) showed a relationship between sit-ups and injury. This research suggests that there is some evidence that better performance on the sit-up test may be associated with decreased risk of injury. Additional evidence suggests that core muscle fitness may reduce falls in elderly populations for both men and women, but that has limited applicability to the military (Granacher et al., 2013).

There is less research on the relationship between muscular strength and injury risk. The reviewed literature demonstrated mixed findings on whether upper- or lower-body strength was significantly correlated with injuries and, if so, whether strength and injury had a positive or negative relationship (Hoffman et al., 1999; Knapick et al., 2001; de la Motte et al., 2017). Overall, research has yet to definitively identify clear relationships between muscular strength and injury for any region of the body.

Evaluation of Specific Measures

Upper Body

Push-ups were one of the most commonly recommended measures of upper-body endurance within the reviewed literature (Hauschild et al., 2014; Palmer and Soest, 1997; Pate, Oria, and Pillsbury, 2012; Suni et al., 1996; Stodden, Sacko, and Nesbitt, 2017; Vogel, 1985). Push-ups have no equipment requirements and good test-retest reliability, $r = .76–.88$ (Hauschild et al., 2014). One workshop SME discussed a prior analysis of AF basic military trainees who were unable to perform a single push-up had a very high risk of attrition before completing basic training.

Pull-ups were another frequently recommended measure (Palmer and Soest, 1997; Pate, Oria, and Pillsbury, 2012; Stodden, Sacko, and Nesbitt, 2017; Vogel, 1985). Pull-ups also have good test-retest reliability, $r = .88–.95$ (Hauschild et al., 2014), but they require more equipment than push-ups (i.e., a pull-up bar). Additionally, one study found that, among British army officers, women performed very few pull-ups (women: Mean = 0.7–2.1, men: Mean = 8.3–10.6; Harwood, Rayson, and Nevill, 1999). The authors concluded that the pull-up test may have been measuring upper-body strength in women while measuring upper-body endurance in men (Harwood, Rayson, and Nevill, 1999); therefore, depending on the physical fitness of the population, it is possible that pull-ups may measure different constructs for men and women.

Handgrip strength is often used as a test of upper-body strength because it requires less equipment than many other tests of upper-body strength (Hauschild et al., 2014; Leong et al., 2015; Pate, Oria, and Pillsbury, 2012; Stodden, Sacko, and Nesbitt, 2017). Nonetheless, a hand dynamometer is necessary for grip strength testing, and this equipment need may reduce the practicality of handgrip strength as a widescale test in military populations. Grip strength has good test-retest reliability, $r = .75–.95$ (Hauschild et al., 2014).

Workshop SMEs discussed a variety of upper-body fitness tests, including pull-ups, push-ups, and the medicine ball toss. SMEs raised some concerns about how well women would perform on certain tests, such as pull-ups, and that these tests might not be a sensitive measure for identifying different levels of muscular fitness at the lower end. In other words, there may be varying levels of fitness among individuals who cannot perform any pull-ups. This concern could be further evaluated to determine whether pull-ups are equally predictive of outcomes for men and women.

Core

Many studies on core endurance recommend the sit-up test as an assessment of muscular fitness (Hauschild et al., 2014; Palmer and Soest, 1997; Pate, Oria, and Pillsbury, 2012; Suni et al., 1996; Stodden, Sacko, and Nesbitt, 2017; Vogel, 1985). Sit-ups require no equipment but have only moderate reliability, $r = .57–.77$ (Hauschild et al., 2014; Courtright et al., 2013). Other measures of core endurance, such as planks, have received minimal attention within the literature but have been recommended as an effective alternative to sit-ups and were identified by at least one SME as a top one or top two replacement for sit-ups. Another test discussed by workshop SMEs was the cross-knee crunch, but more research is needed to evaluate reliability and validity. Because fewer studies have assessed core strength, most assessments often require weight machines (Prieske, Muehlbauer, and Granacher, 2016). More research is needed to determine field expedient tests for core strength.

Lower Body

Timed sprints, typically 100 to 400 meters, correlate well with other lower-body endurance measures (Vogel, 1985; Hauschild et al., 2014). Sprints are field expedient, with no equipment requirements, and also measure speed and power. Sprints of 30 to 600 yards have good reliability, $r = .87–.98$ (Hauschild et al., 2014). However, as noted by one workshop SME, sprints may contribute to more injuries.

The SLJ, also known as the standing broad jump, is another assessment that may be used to address muscular fitness. The SLJ involves measuring the distance a person can travel in a jump that begins with feet next to each other on the same line. The SLJ correlates strongly ($r = .81$) with the one-repetition maximum squat (Koch et al., 2003), and it has minimal equipment requirements and good reliability, $r = .76–.98$ (Hauschild et al., 2014).

The vertical jump, typically starting either from a squat or standing with feet next to each other, is another measure of lower-body strength and power. It correlates well with the one-repetition maximum squat and has good reliability, $r = .80–.98$ (Hauschild et al., 2014; Carlock et al., 2004; Markovic et al., 2004). Different methods are used to measure vertical jump performance, such as measuring acceleration, height, or time in the air. The most accurate measurement appears to be time in the air, assessed with a special jump mat (Aragón, 2000). The increased equipment requirements concomitant with these measurement concerns might reduce the vertical jump's practicality as a field measure.

Potential Limitations

The general health benefits associated with muscular fitness support the inclusion of muscular fitness measures in physical fitness testing batteries. Data from the physical requirements of military task performance also suggest the relevance of upper- and lower-body endurance and strength. Research on injury prevention further suggests that upper-body endurance (i.e., push-ups) and possibly core endurance (i.e., sit-ups) are important for men. The available data for women are less convincing and will require additional research. Gender differences was an issue raised by workshop SMEs for a few different muscular fitness tests, including pull-ups and the leg tuck currently included in the new Army Combat Fitness Test.

Workshop SMEs also raised concerns about the objectivity of scoring sit-ups and push-ups (e.g., what is a good push-up?). This concern is further supported by research indicating that sit-up tests also only have moderate reliability (Hauschild et al., 2014). Workshop SMEs also suggested that airmen sometimes use poor form and technique when performing sit-ups and push-ups. One SME suggested that adding a metronome to control the pacing of the test could partially address the concerns about the test's reliability and airmen's poor form. For example, test-takers would continue to complete push-ups at a prescribed rate until they no longer could maintain the specified cadence. This protocol was used by Yang et al. (2019) in their study of firefighters discussed in a previous section.

Workshop SMEs and the research literature review identified that sit-ups may increase the risk of injury because of the high levels of strain placed on the lumbar spine (McGill, 1995). The current AF protocol of allowing feet to be hooked or held was also raised as a problem for injury risk. However, the extent to which sit-ups and the current protocol contribute to injuries is unclear.

Another limitation may lie in the prioritization of field expediency over other test qualities. Specifically, the jump tests suggested as a measure of muscular strength largely assess power.

While power is a dimension of strength, it does not encompass all elements of strength. This level of specificity is not inherently disadvantageous; in fact, measuring muscular power rather than overall muscular strength reduces the magnitude of sex differences in performance (Courtright et al., 2013). It is, however, worth noting that jump tests may not be the best measure of total lower-body strength, and a workshop SME suggested that it would be impractical to test many airmen on a jump test. Although other tests might better measure lower-body strength, they may also be impractical to conduct in a field setting (e.g., leg press, squat) because of the need for additional oversight from a test administrator. One workshop SME noted that squats were a more functional exercise compared with the current AF sit-up and push-up tests but that the benefits of a squat may be greater for a predeployment assessment than for a Tier 1 assessment for all airmen.

Summary

Muscular fitness is necessary to perform physically demanding tasks. There are various assessments, but the validity of each assessment will depend on the physical demands required by the tasks being performed. There has been considerably less research examining the relationships between muscular assessments and health outcomes, especially in military-aged populations. More research is also needed to address which, if any, muscular FAs are consistently associated with injury risk. Although some alternative FAs could be considered as part of a Tier 1 FA, there is insufficient evidence at this time to suggest that other muscular fitness tests will have stronger validity than push-ups or sit-ups.

Flexibility

Flexibility encompasses the possible range of motion for a certain combination of one or more connective tissues, muscles, and joints (ACSM, 2017; Stodden, Sacko, and Nesbitt, 2017).

Types of Assessments

Tests of flexibility have traditionally focused on static, absolute flexibility. Similar to muscular fitness, flexibility assessments are specific to a given body region. Laboratory tests of flexibility typically measure the angle that a joint can bend, whereas field measures include such tests as the sit-and-reach measure extensibility (Pate, Oria, and Pillsbury, 2012). Although flexibility is not measured as part of the AF-FA, we include it in our review because it is one of the core components of health-related fitness recognized by the ACSM.

Evidence for Inclusion

Limited evidence supports the inclusion of flexibility in physical FAs. In a recent review of flexibility, one study presents the argument that flexibility should even be retired as a major component of physical fitness (Nuzzo, 2019). Overall, findings on the health benefits of flexibility have been equivocal (Stodden, Sacko, and Nesbitt, 2017; Palmer and Soest, 1997; Pate, Oria, and Pillsbury, 2012).

General Health Benefits

Few studies specifically examine flexibility and health. It can be difficult to draw concrete conclusions because many field tests do not uniformly measure the flexibility of a single region. For example, the sit-and-reach test may be a better measure of hamstring or hip joint flexibility depending on each person's final stretch position (Pate, Oria, and Pillsbury, 2012).

Performance of Military Tasks

Similarly, relatively few studies have evaluated the relevance of flexibility for military task performance. Within the studies that have examined this relationship, flexibility has demonstrated only weak correlations with military tasks (Hauschild et al., 2014). Overall, the existing research on the merits of including flexibility in a physical fitness battery is scant and inconclusive.

Injury Risks

The evidence regarding flexibility and injury is also unclear. There is some evidence that flexibility reduces falls in elderly populations, but this has limited applicability for general military populations (Palmer and Soest, 1997). Studies have linked flexibility to lower-back pain or musculoskeletal injury, but many have demonstrated no relationship between these factors (Pate, Oria, and Pillsbury, 2012; Palmer and Soest, 1997). Some studies suggest that very low or very high flexibility are risk factors for injury (Jones et al., 1993; Knapick et al., 2001), but, again, there is insufficient consensus within the literature to draw a conclusion (Pate, Oria, and Pillsbury, 2012). Preliminary research shows that the Functional Movement Screen (FMS), which is a battery of basic flexibility and range-of-motion tests (described in additional detail later), may be associated with injury in military populations, but more research is needed (Lisman et al., 2017; Pate, Oria, and Pillsbury, 2012).

Evaluation of Specific Measures

The sit-and-reach test is the most commonly studied assessment of flexibility (Hauschild et al., 2014). It has low equipment requirements and very high reliability, $r = .80–.99$, but only moderate validity with lab measures of hamstring (.6–.73) or lower-back (.27–.3) flexibility (Pate, Oria, and Pillsbury, 2012). Other suggested measures include the side bend, shoulder stretch, and trunk lift (Suni et al., 1996; Stodden, Sacko, and Nesbitt, 2017; Pate, Oria, and Pillsbury, 2012). A few studies also assessed the FMS (Lisman et al., 2017; Pate, Oria, and Pillsbury, 2012), a test in which participants are asked to successfully and painlessly complete a squat, hurdle step, forward lunge, shoulder mobility test, active straight leg raise, push-up, and rotary stability test. The existing research on these measures is limited.

Potential Limitations

The research reviewed does not definitively provide evidence for the inclusion of flexibility tests on the basis of health benefits, task performance, or injury risk; however, this conclusion is limited by the amount of research on the subject. The role of flexibility in fitness is understudied, and most research examining flexibility focuses on the sit-and-reach test. There is potentially promising research emerging in favor of the FMS, but additional research on the validity, practicality, and fairness of the FMS and other tests is needed.

Additional Perspectives from Subject-Matter Expert Workshops

During the workshops, SMEs discussed other topics related to FA. These topics provide broader context on the policy and implementation issues that the AF could consider in using the FA. We organize the description of SME perspectives according to the key segments of the workshop agenda. Within each segment, we describe the themes identified in our review of the workshop discussion notes. See Appendix A for the workshop agenda, our approach to identifying themes, and the themes we identified in our review.

Health Promotion

When discussing safety and injury prevention at the workshops, participants raised broader concerns about what we would describe as falling within the topic of health promotion. *Health promotion* is "the process of enabling people to increase control over, and to improve their health" (World Health Organization, undated). Several workshop participants raised concerns about airmen not having enough training and resources to prepare for the FAs and to remain physically fit throughout their AF tenure. Several comments were made about airmen engaging in unsafe exercise behavior while preparing for FAs. One participant specifically commented on airmen not getting consistent training at their units and, therefore, starting training only a week or two prior to testing, which can result in injuries. Some commented on the improper technique used during FAs as additional evidence of improper airmen training and also evidence of test administrator training or deficiencies in expertise to teach airmen how to train safely and effectively.

Several participants noted the need to instill healthy lifestyle changes among airmen, especially early in their careers, and the need for dedicated training and resources. Some argued that having a regular FA would act as a motivator for airmen to stay fit. Some also argued for having more resources to train physical training leaders (i.e., the individuals certified to provide physical training and administer the FAs at their units), who would not only ensure test administration but also have the knowledge needed to promote exercise and fitness in their units. More generally, a participant noted that that the AF (unlike the Army) does not have offices dedicated to providing evidence-based health promotion for airmen. This person noted that this has resulted in an ad hoc system in which some commands may be funding health resources while others are not.

Guard and Reserve

Two general themes came up in discussions about Guard and Reserve FAs: (1) resources for testing and (2) frequency of test administration. In terms of resources, some participants raised the challenge of not having enough test administrators and limitations with access to facilities. For example, one participant noted that there are not enough physical training leaders to observe each member of the unit perform sit-ups and push-ups. In terms of testing frequency, participants commented that more-frequent fitness testing might incentivize Guard and Reserve members to build their fitness and might ensure that more of them take the test before they have to drill. For example, one participant proposed the idea of 12-month, 18-month, and 20-month testing schedules but with the incentive that, if a member performed well at 12 months, the member would earn the right not to test again for another 18 months (and so on). At least one participant did warn that a regular testing schedule during cold-weather months (e.g., January) at units in colder climates would require alternatives to outdoor tests (e.g., using a beep test indoors instead of 1.5-mile run that is typically performed outdoors). In general, participants who discussed Guard and Reserve issues expressed preferences for more-regular fitness testing and more resources (or alternative tests) to ensure that the testing is completed in a more standardized manner across units than what is currently in place.

Other Themes

We noted other policy-related themes, some of which emerged during the workshops' "special topics" segment.

Database and Data Quality

Some participants noted challenges with using the AF Fitness Management System (AFFMS).[11] A few noted that airmen are inappropriately tagged as having a medical exemption in AFFMS when they should not be because those airmen had taken an alternative test that the system did not capture. One participant noted challenges with receiving data in a timely manner, indicating that it can take up to two weeks. In general, participants who raised the topic of AFFMS did not find that it was flexible and questioned its value.

[11] AFFMS is on its second version, known as AFFMS II. According to an online post from Air Force Personnel Center Public Affairs, AFFMS II "improves accessibility and fitness program managers' ability to manage fitness program records. In addition, it features more stringent security controls to protect members' information from unauthorized changes" (Gildea, 2015).

Policies to Incentivize Fitness

Random testing (i.e., lack of predictable FA schedule) was cited as a way to ensure that airmen do not "cram" for their FAs and are motivated to stay fit. One participant also suggested the alternative of randomizing which tests (e.g., beep test versus 1.5-mile run) airmen receive at each assessment period to encourage airmen to train using different exercises.

Some participants suggested that if airmen perform well on their FAs, they can be tested less frequently. However, some participants raised concerns about this idea being applied to Guard and Reserves and also for deployment. Others offered suggestions on other possible incentives, such as bonuses, time off, promotion points, and others. One participant raised the question of whether airmen could be given a set of incentives from which they could choose. Yet other ideas focused on having fitness competitions (e.g., by major command), aided by online applications.

Concerns About Exemptions and Waivers

Threaded throughout the discussion about specific fitness components and associated tests were issues about waivers and exemptions. For example, cold weather at some base locations results in more weather-related waivers for outdoor tests. Another area of discussion about waivers and exemptions concerns the feasibility of using certain tests (e.g., row ergometer) that could contribute to medical waivers because of the physical demands for both the upper and lower body (e.g., airmen with shoulder limitations could get exempted from the row ergometer but not necessarily from the 1.5-mile run).

Another concern about waivers and exemptions involves their use to inappropriately bypass testing (i.e., "gaming the system"). This concern was raised in both the context of asking whether all medical personnel assessing airmen medical fitness have sufficient understanding and skill at determining airmen eligibility for fitness testing and asking whether airmen have "skill" at finding loopholes in policies that allow them to get exemptions or waivers from all or some parts of the FA. Gaming through use of waivers and exemptions was of particular concern in the context of a potential predeployment assessment, raising the question of whether some airmen perform poorly on a Tier 1-D assessment to get out of deploying.

Miscellaneous Topics

Several other policy-related topics were raised, but none rose to the level of a theme. We list some of these topics as follows:

- *Recruiting and accessions:* incentivizing recruiters for airmen who graduate from training and have preaccession fitness standards beyond the medical standards currently in use
- *DoD policy:* updating AF policy may be restricted by current DoD policy that specifies which fitness components need to be measured
- *Age-based testing scores:* concern about the size of the age group categories used for scoring each component (i.e., using same fitness test scores for older members as for somewhat younger members)

- *Compliance:* question of whether inspections (via inspector general function) be used to ensure that people are accurately self-reporting fitness information
- *Fitness versus testing:* moving away from focus on fitness testing to focus on ensuring that airmen exercise regularly.

Summary of Health-Related Fitness Components

The research and discussions with workshop SMEs show that there is strong consensus that the AF is measuring important fitness components that are related to a wide variety of health outcomes. The evidence also suggests that the AF assessments are generally effective measures. In particular, there is strong evidence supporting the 1.5-mile run and the AC as indicators of health risk. Although sit-ups and push-ups measure muscular fitness, there were concerns from workshop SMEs that these tests can be subjective and may not produce very reliable or accurate scores. Several alternative tests were considered to address these and other potential gaps in the current AF-FA, which will be discussed in more detail in Chapter 4.

One important question not addressed by existing research is the optimal number and combination of assessments that should be used by the AF to predict health risk or promote a fitness culture. If assessments are highly correlated, the AF may be able to use fewer assessments and obtain similar estimates of health risk. However, minimizing or reducing the number of assessments could have unintended consequences on the AF's goal to promote a fitness culture. For example, airmen may associate reductions in FAs as signs that leadership does not fully support or value fitness. More research is clearly needed to address these potential trade-offs and potential strategies that use minimal resources but also promote a fitness culture.

3. Fitness to Deploy and Other Policy Considerations for Physical Fitness

This chapter provides an analysis of fitness-related training and requirements related to an airman's fitness to deploy. In particular, we identify physically demanding tasks performed in deployment readiness training courses and evaluate whether the AF-FA sufficiently covers the abilities required to perform these tasks. There is currently no other general AF-wide FA—therefore, we consider the AF-FA as a baseline to evaluate predeployment fitness even though the AF-FA is primarily designed to promote general health. If deployment tasks require physical demands not currently covered by the AF-FA, the AF might need to explore developing a Tier 1-D assessment.

Overview of Deployment Policies

The principal documents providing the structure for deployment readiness are provided in AFI 10-405, *Expeditionary Readiness Training* (2018), and AFI 48-123, *Medical Examinations and Standards* (2013). The AF organizes deployment environments and corresponding training requirements into three categories: (1) basic airman readiness, (2) basic deployment readiness, and (3) advanced deployment readiness (see Figure 3.1). To be considered deployable, all airmen must meet certain basic requirements, some of which are related to physical readiness:

- must be able to perform duties of AFSC for a prolonged period (at least 12 hours)
- at a minimum, must be able to carry all required deployment baggage (at least 40 pounds)
- must be able to run at least 100 yards to take cover
- must be able to perform duties in hot and cold environments.[12]

In addition to these requirements, basic airman readiness training is designed to "establish/maintain a strong warrior ethos while also providing them a solid foundation for knowledge, skills and abilities that prepare them to survive, operate and succeed across the full range of military operations" (AFI 10-405, 2018, p. 10). However, discussions with representatives from the Expeditionary Readiness Center noted that basic airman and basic deployment readiness training largely consist of computer-based training and other nonphysical training requirements. We conducted a limited number of interviews ($n = 12$) with airmen to identify physical demands associated with basic deployments (see Appendixes B and C). These tasks generally included some lifting, carrying, bending, and walking as airmen perform their duties (e.g., move equipment and supplies).

[12] For more details, see AF Medical Support Agency, 2018.

Discussions with policy representatives from Headquarters, AF (A1 and A3) and Air Education and Training Command suggested that most predeployment physical demands would be experienced in advanced deployment readiness courses designed to prepare airmen primarily for uncertain and hostile environments. These courses are described in more detail in the following section.

Figure 3.1. Air Force Expeditionary Readiness Training

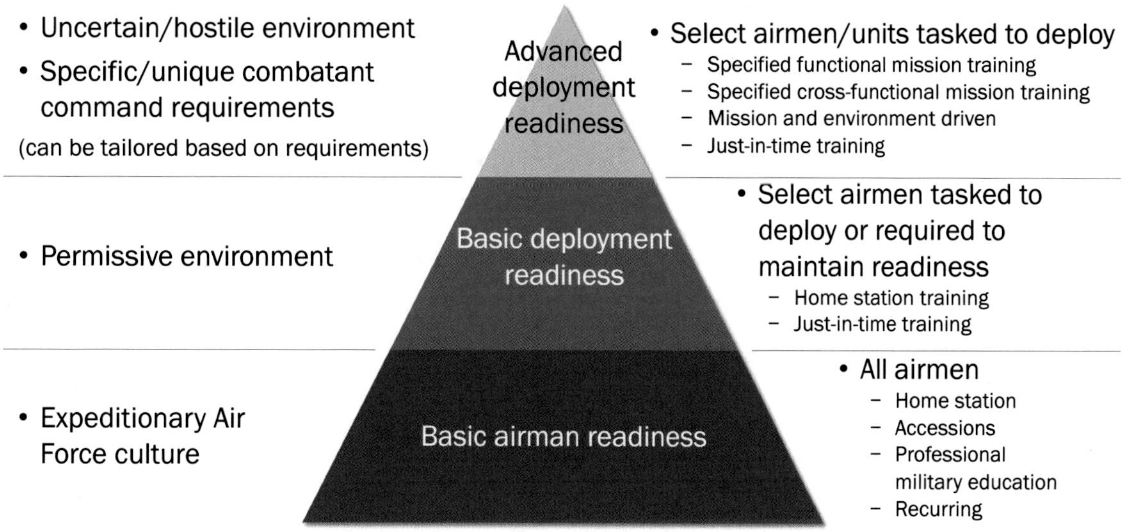

SOURCE: AFI 10-405, 2018.

Fieldcraft Training Courses

In the following sections, we provide a brief description of fieldcraft (FC) training courses followed by an analysis of the physical components and demands associated with these courses. Our analysis is based on discussions with FC instructors and leadership, observations of airmen completing selected portions of the training focused on the physical demands, and brief focus groups with airmen in training at the 421st Combat Training Squadron at Joint Base McGuire-Dix-Lakehurst in New Jersey. We also conducted a systematic evaluation of the physical demands associated with training using ratings provided by FC course instructors and personnel from the AF ESU.

Background and Historical Attendance

Advanced deployment training is provided across three primary FC courses: (1) Fieldcraft Hostile (FC-H), (2) Fieldcraft Uncertain (FC-U), and (3) Fieldcraft Contingency Response

(FC-CR).[13] Started in 2012, these FC courses were designed to replace the Combat Airman Skills Training course that continued through 2015. FC courses provide training to prepare airmen with a variety of skills needed to operate in environments that are overtly hostile (FC-H), where security is unknown or limited (FC-U and FC-CR). These courses are cross-functional and include airmen from a broad variety of specialties. The topics cover "knowledge, skills and abilities not routinely acquired during occupational and/or other ancillary training venues" (AFI 10-405, 2018, p. 18).

Attendance may be required because of several factors, including "specialty training, deployment location, threat assessment, specific mission, duty assignment, role, operation, or special requirement" (AFI 10-405, 2018, p. 11).

A review of FC course graduation data (Figure 3.2) shows that the number of airmen attending advanced deployment training is very small compared with the total number of active-duty airmen.[14] Although attendance at the FC courses has gradually increased each year since 2012, there were still fewer than 5,000 airmen graduating from all FC courses in 2019. Other physically demanding predeployment training is designed for airmen within a specific AFSC (e.g., security forces). We did not include this training as part of our evaluation because it did not apply to the general AF population.

---

[13] A fourth course, FC CENTCOM, was also offered between 2012 and 2019 but has since been phased out.

[14] As of January 1, 2019, there are more than 300,000 active-duty personnel in the AF (Air Force Personnel Center, 2020).

Figure 3.2. Number of Airmen Attending Advanced Deployment Training Courses

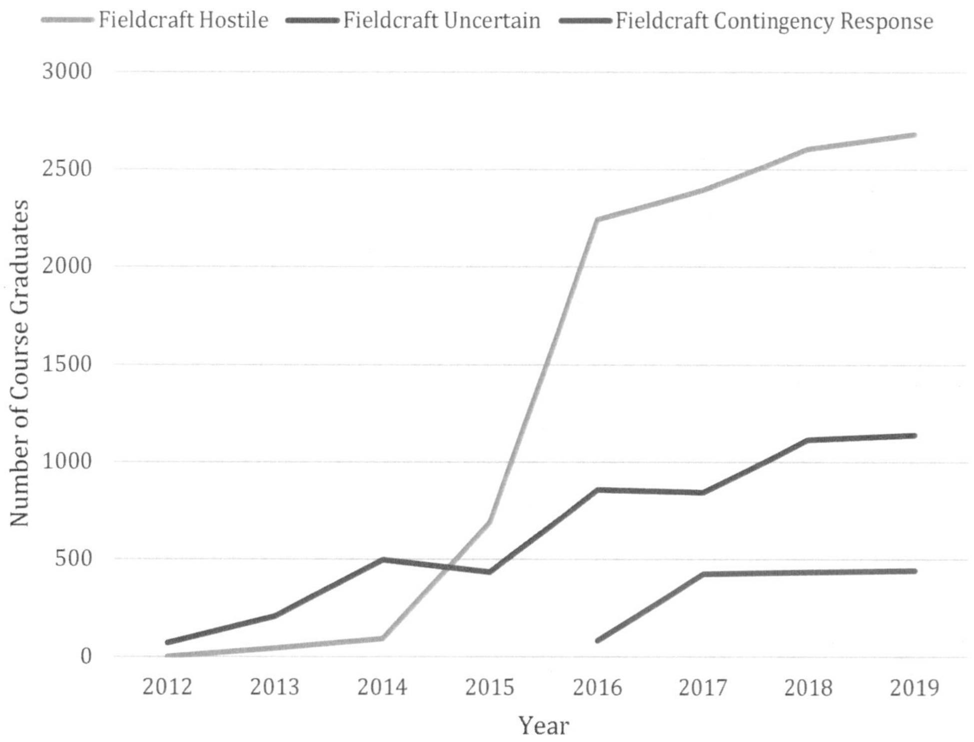

Instructor Discussions and RAND Observations of Training

The research team conducted semistructured interviews with predeployment training instructors at Fort Dix, New Jersey. We interviewed eight instructors who taught a variety of predeployment courses (i.e., FC-H, FC-U, FC-CR). Overall, there was a consensus among instructors that not all airmen are physically prepared for predeployment training when they arrive, and they are not all prepared to meet the physical demands of deployment when they leave.

Instructors reported that a notable number of airmen lack the cardiovascular and muscular fitness required for training, which is compounded by the fact that many airmen are not used to performing activities wearing full gear (i.e., body armor, helmet, water, rifle). Several trainees confirmed that wearing and moving on foot with full gear for several hours was physically demanding, especially in the heat.

In training, insufficient physical fitness appears to prevent airmen from completing required exercises or causes them to sustain injury while doing so. One instructor stated, "without equivocating," if airmen were fitter prior to training, there would be fewer injuries. Relatedly, instructors also noted that some airmen arrive for training while "on profile" for an injury, meaning that they cannot perform some of the training tasks.

While predeployment training is believed to improve airmen's physical fitness, all instructors said that some airmen are unable to reach a level of fitness necessary for deployment by the end of the course. There is no formal process in place for instructors to receive feedback about the physical preparedness of airmen once deployed; however, informal feedback suggests that not all deployed airmen have adequate physical fitness to perform their duties in their deployed field environment. Consequently, fitter airmen may have to take on a disproportionate share of the work.

In addition to discussions with instructors, we observed several training events that involved reacting to an ambush during a dismounted movement, urban tactics and team movements, hand-to-hand combat, casualty evacuations (i.e., litter carry), and direct fire during a mounted movement. Each of these scenarios included several physically demanding movements, to include sprints and quick movements (about 20 to 150 meters), getting into a prone position to take cover, crawling, long marches (about 6 miles) on foot while wearing body armor and rucksack, lifting and carrying a casualty, jumping into a vehicle, and grappling with an enemy combatant.

Although most trainees we observed performed the training tasks, a few airmen sometimes could not complete a task without unscheduled rest breaks. Instructors noted that such rest breaks could be catastrophic during a deployment depending on the situation (e.g., airmen who need to catch their breath and recover while moving a critical casualty from the field of combat to an evacuation point). Trainees also noted that completing multiple casualty movements was particularly challenging.

Analysis of Physically Demanding Tasks

To more systematically evaluate the specific demands associated with the FC training courses, we asked nine training instructors (three from each course) to identify the movement patterns required to perform training tasks listed in official course documents.

Instructor Ratings

We used a five-step process to identify physical demands associated with FC training courses:

1. We extracted 142 training tasks from the FC training course plans of instruction.
2. We completed an initial sorting of the training tasks into physical and nonphysical demands. Of the 142 training tasks, 31 were categorized as physical (i.e., tasks requiring moderate to high levels of physical effort) and 111 tasks were categorized as nonphysical (i.e., requiring low to no physical effort). All training tasks were listed in a Microsoft Excel workbook, with one spreadsheet containing the physical tasks and a second spreadsheet containing the nonphysical tasks.
3. For each physical task, instructors independently provided two sets of ratings. The first rating indicated the level of physical effort required to perform the task using a scale

rating from 1 (Very, Very Light) to 7 (Very, Very Hard). Second, instructors rated whether a specific movement pattern (e.g., squat, lift, run) was required by placing an "X" in the appropriate cell.

4. Training instructors were also asked to review the nonphysical tasks to confirm our initial sorting into physical and nonphysical categories. We asked instructors to provide ratings only for any "miscategorized" tasks that required physical effort. Four tasks on the nonphysical task list received an average rating of 4 (Light) or higher. These included (1) "apply Tactical Combat Casualty Care (TCCC) in an uncertain environment," (2) "perform casualty care using TCCC," (3) "Supported casualty care (US MEDIVAC)," and (4) "nonsupported casualty care." These four tasks focused on casualty care were subsequently moved to the physical task list.
5. Instructor ratings were aggregated, and tasks with overlapping content were consolidated to broader task activities (e.g., combat casualty care). These task activities are presented hierarchically in a heat map (Figure 3.3) with task activities at the top that required, on average, the most physical effort and movement patterns. The darker shading indicates a higher percentage of instructors indicating that the task was demanding or that a particular movement pattern was required.

Figure 3.3. Physical Demands Associated with Advanced Deployment Training

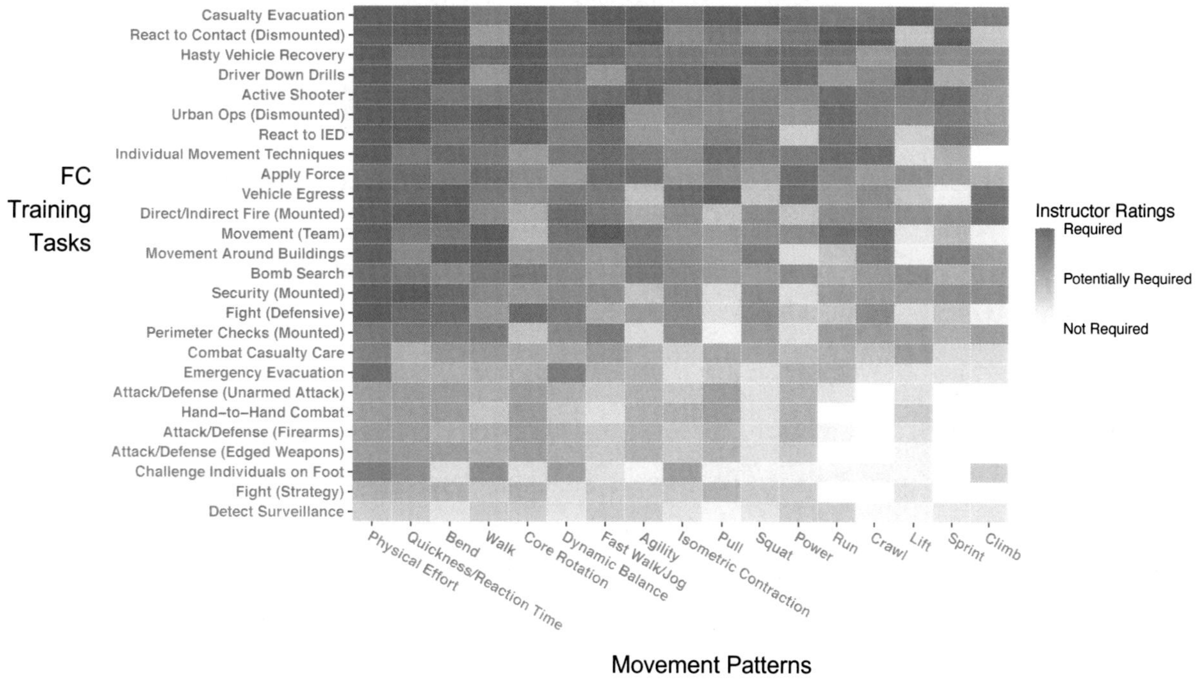

NOTE: ops = operations.

These ratings suggest that advanced deployment training contains tasks that both are physically demanding and require a wide variety of movement patterns to perform. In particular, tasks that involve responding to emergency situations (e.g., casualty evacuation, react to contact) were rated as physically demanding and involve several movement patterns, including quickness, bending, core rotation, and agility. Some training tasks that were rated relatively low in terms of physical effort (e.g., hand-to-hand combat) may require significantly greater physical demands if encountered during a deployment. Follow-on discussions with instructors suggest that some training tasks were more knowledge-based than physical performance–based. Additional analyses may be required to more completely identify the components and level of fitness required to perform tasks in an actual combat environment.

Given the physical demands associated with these and many other advanced deployment training tasks, we considered whether the AF-FA contains the right tests and standards to ensure fitness to perform deployment missions. For example, a task that requires primarily walking and/or running but few other movement patterns could arguably be covered by the 1.5-mile run used as part of the AF-FA. Even if the 1.5-mile run is the right test, the age- and gender-graded standards used to score performance on the AF-FA may not be sufficient to ensure that airmen can perform specific deployment tasks.

To evaluate how well the AF-FA measures the abilities required to perform training tasks in the FC courses, we asked seven ESU personnel to answer four questions:

1. Are the AF-FA tests sufficient for measuring the physical demands of each FC training task?
2. Are the AF-FA performance standards sufficient to ensure capability to perform each FC training task?
3. Which movement patterns (e.g., bend, squat) are not sufficiently addressed by the AF-FA for performing each FC training task?
4. What physical fitness components (e.g., muscular power, agility) need to be measured to address potential gaps in the AF-FA?

Exercise Science Unit Ratings

To answer the first two questions about the adequacy of the AF-FA and the corresponding standards, we computed the percentage of ESU raters indicating "sufficient." If less than two-thirds of the raters indicated that a task was sufficiently covered by the AF-FA, we considered that as a potential gap. Overall, about half (52 percent and 56 percent) of the FC tasks were rated as not sufficiently covered by the AF-FA tests and standards, respectively. In general, tasks that were rated as sufficiently covered by the AF-FA required movement on foot or required less physical effort (e.g., movement around buildings/team movements, detecting surveillance).

To more fully understand the nature of these gaps, ESU raters indicated which movement patterns the AF-FA does not sufficiently cover for each training task. We aggregated these ratings and summarized them in a heat map in Figure 3.4, which illustrates several patterns. First, certain movement patterns (e.g., walk, run, fast walk/jog) are covered by the AF-FA, most likely the 1.5-mile run. Several tasks (i.e., rows in white) also appear to be sufficiently covered by the AF-FA.

Second, the current AF-FA does not appear to be a sufficient assessment for the physical movement patterns required by several FC training tasks. In particular, reacting to contact requires a variety of movements, to include short sprints, getting down into a prone position on the ground, and moving on the ground while maintaining a low profile (e.g., low and high crawls). The ability to perform a casualty evacuation also requires movement patterns not covered by the AF-FA. Casualty evacuations require the ability and power to squat, lift (e.g., litter carry), and pull casualties (e.g., casualty drag) to a safe area.

The figure also highlights several potential gaps (i.e., medium shading)—these are tasks that did not receive consistent evaluations from ESU raters. Feedback from ESU raters suggests that these tasks require further clarification to better describe the actual task being performed. Furthermore, some tasks are defined too broadly. For example, moving as a team (movement [team]) may require different movement patterns depending on the environment, formation, and distance covered. Future analysis is required for these tasks to identify which, if any, parts of these tasks are physically demanding and whether those are different from other tasks already documented.

Figure 3.4. Exercise Science Unit Ratings of Movement Patterns and Fieldcraft Training Tasks Not Sufficiently Covered by the Air Force Fitness Assessment

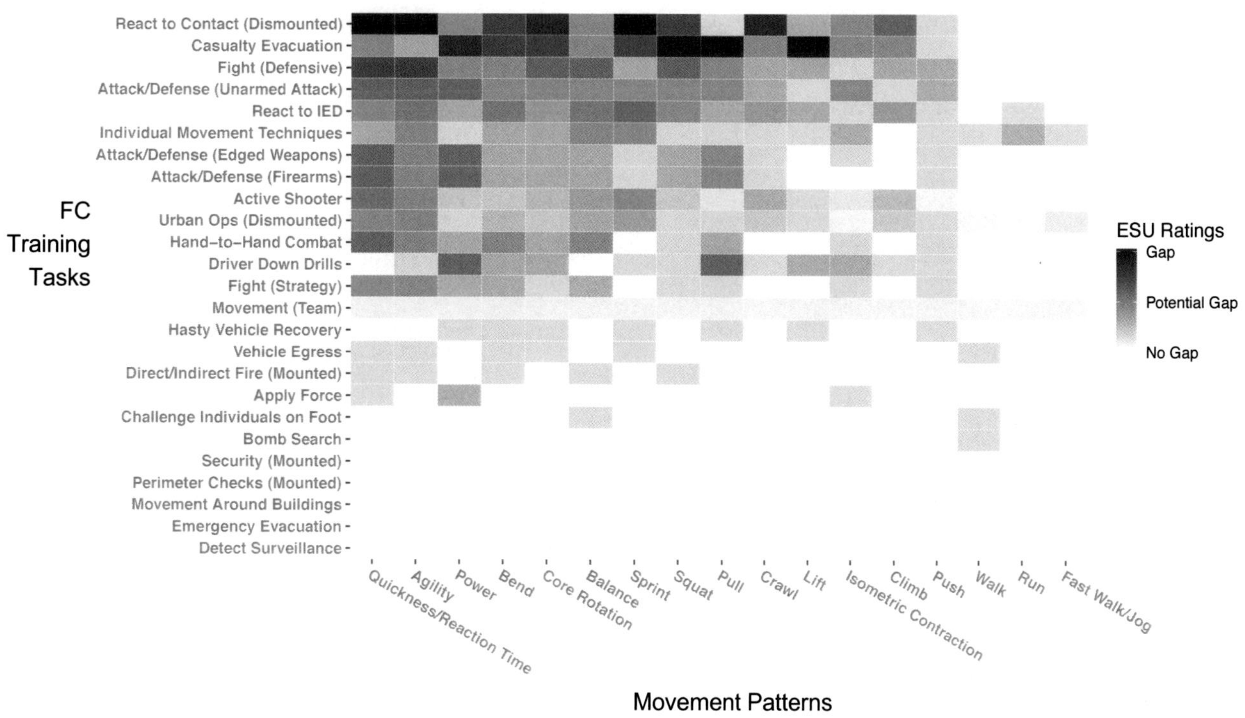

NOTE: ops = operations.

Finally, we asked ESU raters to consider any gaps in the AF-FA and recommend the specific physical fitness components that should be assessed to ensure that airmen are capable of performing the training tasks in advanced deployment training. Because the purpose of this evaluation was to inform an evaluation of fitness for deployment (i.e., measure the abilities to perform physically demanding tasks), several skill-based components of fitness (e.g., reaction time, agility, speed) were added to the components already measured by the AF-FA.[15]

For example, we asked whether agility was recommended as a fitness component to address any gaps in the AF-FA for performing the bomb search training task. We aggregated these yes/no recommendations across tasks and computed percentages for recommended fitness components (Table 3.1). Thirty-one percent of ESU responses across tasks suggest that reaction time is needed to address gaps in the AF-FA.

Table 3.1. Physical Fitness Components Needed to Address Gaps

Component	ESU Rated as Needed
Reaction time	31%
Agility	25%
Balance	25%
Speed	22%
Coordination	20%
Muscular power	19%
Muscular strength	17%
Anaerobic capacity	15%
Flexibility	14%
Muscular endurance	12%
CRE	9%
Body composition	5%

Summary of Fitness to Deploy

Beyond a few basic requirements, the physical demands of deployment readiness are associated with preparations for advanced deployments to environments that are uncertain or hostile. The number of airmen attending these advanced courses is relatively small compared

[15] A predeployment FA would correspond more closely to a Tier 2 test, which is designed to measure physical fitness levels needed to perform specific occupational tasks that correspond with an assigned AFSC. Consequently, the tests and standards in a predeployment FA should be age- and gender-neutral.

with the total population of active-duty airmen. Advanced deployment training can be physically demanding and may require physical fitness components not currently assessed by the AF-FA. Furthermore, airmen attending advanced deployment courses may not always be sufficiently fit to effectively perform the training tasks and may not develop the fitness levels during training to perform the tasks that could be required during a deployment. Currently, there is no systematic assessment of an airman's capability to perform these physically demanding tasks and no accountability or feedback provided to commanders about an airman's physical readiness to perform these tasks. Using these data and observations, the AF might need to develop a Tier 1-D predeployment FA that is administered in addition to the general Tier 1 FA.

4. General Conclusions and Recommended Courses of Action

Overall, the current AF-FA addresses the most-critical fitness components to promote airmen health and minimize injury risks (see Table 2.1). In particular, the 1.5-mile run and the AC are strong assessments supported by a great deal of research. Muscular fitness is primarily required to perform physically demanding tasks. Although the muscular FAs (i.e., sit-ups, push-ups) used by the AF are acceptable measures of muscular endurance, the validity of assessments will depend on the physical demands required by specific tasks. More research is also needed to establish the relationships between muscular fitness and health outcomes, especially in military-aged populations. This line of research is needed to identify which assessments are most important and the minimum standards for those assessments that are associated with reduced health risks. Although flexibility is not currently assessed by the AF, the research is inconclusive on whether certain tests and standards would provide any benefits (e.g., reduce injuries).

The remainder of this chapter will provide an overview of the specific strengths and potential limitations of the current AF-FA. It is important to note that no single test is best for any given physical fitness component. However, some tests may provide some advantages in terms of injury risk, cost, ease of administration, perceived fairness, reliability, and validity. We considered each of these factors when developing recommendations and courses of action that should be considered to further strengthen the position of the AF to ensure its airmen are healthy and physically ready for duty.

Strengths in Current Air Force Fitness Assessment

Cardiorespiratory Fitness and Abdominal Circumference Are Important Indicators of Health Outcomes

There is strong consensus that CRE and body composition are critical to an effective FA that can be used to support general physical health of the force. Although there are many alternative tests that could be used to assess these fitness components, the 1.5-mile run and WC have been consistently supported as valid measures for CRE and abdominal fat (a critical area of body composition). The AF could further strengthen its measure of body composition by adopting the WtHR. This point is further discussed in the recommendations section.

Criterion-Referenced Scores Are Used for the Abdominal Circumference Standard and the 1.5-Mile Run

Our review did not evaluate the specific time cutoffs used by the AF. However, we acknowledge some other positive elements associated with the implementation of the 1.5-mile run and the AC assessments. Discussions with the AF ESU and internal AF documents suggest that the test standards for the 1.5-mile run and AC are both criterion-referenced, meaning that scores are associated with different levels of health risk. Although criterion-referenced cutoffs are strongly recommended, the AF relies on broader population research conducted by external research organizations. The AF is encouraged to conduct research using its own data to establish relevant cutoff scores to ensure that results generalize to all airmen. This recommendation will be discussed in more detail in a later section of this chapter.

The Air Force Fitness Assessment Is Practical

Workshop SMEs noted that the current AF-FA requires minimal test resources and can be administered almost anywhere. Furthermore, testing large groups of airmen is possible because none of the tests require exercise equipment.

Potential Concerns with the Current Air Force Fitness Assessment

With any FA, there will always be some limitations. We summarize concerns with the AF-FA in the following sections.

Potential Risk of Injury Related to Taking or Preparing for the Different Tests

Although there will always be a risk of injury when performing any physical activity, there were increased concerns specifically for sit-ups and running. In particular, concerns about the strain placed on the lower back during sit-ups emerged from the literature review and our workshop SMEs. Prior research suggests that there are several factors that can increase the risk of injury during a sit-up, one of which—allowing feet to be secured or hooked—is related to the current protocol used by the AF. There was also some concern with the high rate of exemptions for the 1.5-mile run that could be related to the high impact of running.

Muscular Fitness Is Multidimensional, But Current Fitness Assessments Only Measure Muscular Endurance for a Subset of Muscle Groups

In a review of abdominal fitness tests, Knudson (1999) noted that "no one abdominal exercise is best and abdominal muscle activation is quite variable," suggesting that multiple assessments may be needed. Muscular strength is also not assessed in the current AF-FA. Although more research is needed to evaluate the relationship between muscular strength and injury prevention, there is some evidence to suggest that muscular strength is related to cardiovascular disease and all-cause mortality (Leong et al., 2015). Furthermore, muscular strength can be an important component for performing common military tasks that may be required during deployment.

Fairness and Validity of the Abdominal Circumference Measure

Some concerns were raised about the face validity and perceived fairness of the AC measurement, as AC is not sensitive to stature. That is, the health risk may be different for airmen who have the same AC but are significantly different in height.

Some Fitness Assessments Are Subjective

Although test administrators are provided with a standardized protocol and instructions, push-ups and sit-ups require more subjectivity to evaluate, which can negatively affect the reliability and validity of the test. Several workshop SMEs noted that airmen sometimes use poor form and technique when completing push-ups and sit-ups. Furthermore, there may not be a common consensus on what "counts" as a good push-up or sit-up. Although training test administrators and airmen on correct form can partially address this concern, evaluating dynamic movements to determine a correct repetition is inherently more challenging compared with timing a 1.5-mile run.

The Current Scoring System Is Not Sensitive to the Relative Importance of CRE

Research has shown that higher levels of CRE may offset some of the health risks associated with body fat. Although changing to a compensatory scoring system may be supported scientifically, the added complexity could increase the difficulty of implementation and lose the ability to easily interpret meaning from scores. Nonetheless, a compensatory scoring system has the potential to be more predictive of health risk compared with simple cutoffs on multiple assessments.

The Air Force Does Not Fully Address the Physical Fitness of Airmen for Advanced Deployments, Specifically to Hostile or Uncertain Environments

Although the AF-FA is designed for promoting health, we began our analysis of deployment fitness by considering whether the AF-FA components could be used and if they would be sufficient. Our preliminary analysis based on ratings provided by training instructors and ESU personnel suggests that some deployment training tasks (e.g., running) may be effectively covered with existing AF-FA tests (e.g., 1.5-mile run). However, several other training tasks require skill-based fitness components (e.g., agility) that are not currently measured by the AF-FA.

Recommendations

To address the potential gaps and limitations with the current AF-FA, we provide several recommendations and courses of action.

Review and Provide Corrective Feedback as Necessary to Reduce Potential Injuries and Improve Airmen's Training, Preparation, Form, and Technique

There is always some risk of injury when performing any physical activity. However, comments from our workshops raise some more-specific concerns, particularly about how airmen prepare for the AF-FA and their physical form when completing certain assessments. As noted during our workshops, airmen may take the AF-FA once or twice a year, so their activities the other 364 days of the year are critical to reinforcing proper training techniques and maintaining good health. If airmen have not been consistently exercising, they may try to prepare by cramming in many repetitions and miles in a short period of time, which may contribute to overuse injuries (Kaufman, Brodine, and Shaffer, 2000).

Conduct a Trial Study to Explore Alternative Assessments and Administration Protocols and Determine the Optimal Number and Combinations of Assessments

We recommend selecting several alternative assessments to further evaluate in a voluntary trial study. A well-designed trial will allow the AF to collect the necessary data to determine whether (1) assessments can be scored reliably, (2) scores are linked to important outcomes, (3) airmen can train and perform the assessment safely, (4) airmen perceive value in the assessments, and (5) assessments are fair to all airmen regardless of demographic characteristics (e.g., age, race, gender).[16]

As part of this study, we further recommend conducting a cost-benefit analysis prior to adopting any changes to ensure that resources will be used effectively. This analysis will help address how many assessments are needed to meet intended Tier 1 or Tier 1-D objectives. Because FAs are often correlated, there will be a point of diminishing returns at which adding more assessments will yield little to no incremental validity. To guide these decisions, the AF should establish metrics or benchmarks for acceptable levels on specific outcomes (e.g., diabetes risk). If the level of risk is already acceptable, the AF may decide that additional (or alternative) assessments are not worth the increased resources required to train, administer, score, and manage. Although there is a vast number of potential alternative tests to consider, we offer a few to consider based on a combination of the research literature review and discussions with workshop SMEs.

---

[16] Fairness is a multifaceted construct and can have different meanings depending on the context. For a Tier 1 health-based assessment, a single standard shared by two subgroups can be considered fair when the health risk is the same for a specific outcome. For example, to achieve similar health risks, the AC standard for women would need to be thinner than the AC standard for men. For a Tier 2 or Tier 1-D standard, a standard is fair when applied equally to all subgroups and there is a lack of predictive bias, meaning that the fitness test predicts task performance equally well across subgroups. For a more thorough discussion of fairness and test bias, see Society of Industrial and Organizational Psychology, 2018.

Planks as an Alternative to Sit-Ups to Measure Core Stability

Considering the arguments suggesting that there may be potential injury risks associated with sit-ups, planks may offer an alternative that places less strain on the lower back but still recruits different core muscle groups (Harvard Health Publishing, 2020). When considering potential alternatives to the sit-ups for the Navy, Peterson (2013) concluded that the "plank is an easy test to administer, operationally applicable, and may actually reduce the likelihood for and occurrences of low back injuries if trained regularly" (p. 26). More research is needed, however, to determine whether planks or any other core assessment will reduce injury risks. At this time, these claims are based on very limited research. One drawback to using planks is the potential for test administration to take longer than sit-ups. Research with college students found that the median test duration was 72 seconds and 110 seconds for women and men, respectively (Strand et al., 2014). Additional research is also needed to determine whether there is a specific amount of time that is associated with better health and injury outcomes.

Other research comparing musculoskeletal injuries among soldiers randomly assigned to either a core stabilization training program (e.g., planks) or a traditional exercise program found some evidence in favor of core stabilization exercises (Childs et al., 2010). However, in a comparison of several exercises, Axler and McGill (1997) found that "no single exercise best recruited all of the abdominal muscles simultaneously." Clearly, more research is needed to evaluate different core exercises and assessments and their relationship to important outcomes (e.g., fewer injuries) among military personnel.

Squats, Deadlift, and Lunges as Measures of Muscular Strength and Endurance

Although we recommend planks as an alternative to sit-ups, conclusions drawn from a systematic review suggest that squats, deadlifts, and other free-weight exercises can effectively train the core while also targeting other muscle groups (Martuscello et al., 2013). Therefore, it is possible that the AF may not need to include a specific core measure if multijoint free-weight assessments are used. However, the inclusion of assessments with weights requires equipment and additional training and education for both test administrators (e.g., Physical Training Leader) and airmen to mitigate any potential increases in the injury risk.

Cadence to Control Rate of Repetitions

Prescribing a set pace for a muscular fitness test (e.g., 25 repetitions per minute) would allow physical training leaders to emphasize form and technique over pace. In a test using cadence, airmen would continue to perform repetitions until they could no longer maintain the specified cadence controlled by a metronome. Advocates of cadence for sit-ups suggest that a standardized pace can "eliminate potentially dangerous bouncing and ballistic actions in the hips and back" (Knudson, 1999, p. 50). However, completing repetitions to a prescribed cadence rather than self-paced is more difficult—therefore, additional training and practice may be required to learn how to perform correctly (McGuire et al., 2011).

Waist-to-Height Ratio as a Measure of Body Composition

Finally, the AF could consider a measure of body composition that incorporates body height. As previously discussed, the WtHR and WC are effective predictors of health outcomes. The primary advantage of WtHR is that a single cutoff could be used for both men and women. The current AC cutoff is 39 inches for men and 35 inches for women. If a WtHR were adopted, the AF could use a single cutoff (e.g., .5) that would be applied to both men and women.

Use Air Force Data to Evaluate the Benefits of a Single Index That Integrates Scores from Body Composition and Cardiorespiratory Endurance

Although the 1.5-mile run and the AC assessment are both valid and effective measures of fitness, the AF could consider developing a single index that accounts for the interrelationships between CRE, body composition, and health outcomes. In addition to recent evidence suggesting that such indexes (FFI) are potentially more predictive of health outcomes, a single index may increase perceptions of fairness by addressing the complex relationships between these fitness components (e.g., "fit but fat"). Although the FFI as originally developed would require additional equipment (a treadmill), the AF could conduct additional research to develop a measure that is similar to the FFI but uses run times from the 1.5-mile run test to calculate an index incorporating both body composition and cardiorespiratory fitness.

Establish Criterion-Referenced Cutoff Scores All Fitness Assessments

Based on discussions with the ESU, only the 1.5-mile run and the AC assessment use criterion-referenced cutoff scores. We strongly recommend establishing criterion-referenced scores for all components of the AF-FA (e.g., push-ups, sit-ups, other muscular fitness test). Although this requires additional research, the AF maintains some of the largest databases containing fitness scores, health outcomes, and performance outcomes. Linking relevant data would allow the AF to establish meaningful cutoff scores directly tied to AF-FA objectives. That is, the AF could determine how many push-ups and sit-ups are needed to decrease the risk of injury, improve health outcomes, and be physically fit for deployment.

Ensure Appropriate Use of Gender-Specific and Gender-Neutral Standards

Standards for a predeployment fitness test should be age- and gender-neutral (e.g., single standard on an agility test) to ensure that everyone can perform the required tasks. However, standards for a health-based FA may need to vary based on age and gender to ensure that standards are tied to the same level of health or injury risk (Hardison, Hosek, and Bird, 2018). For example, the percentage of body fat that is considered healthy is different for men and women (ACSM, 2018). However, to the extent supported by the data, the AF should strive to implement a single standard for each Tier 1 assessment to further promote perceptions of fairness and acceptability.

To fully evaluate the appropriateness of either Tier 1 or Tier 1-D standards, the AF will need to conduct predictive bias analyses across different subgroups (e.g., gender, age, race) using existing data or by collecting additional data. To conduct these analyses, the AF would first need to ensure that it has assessment data and outcome data (e.g., injuries, medical diagnoses) for a sufficient number of airmen representing all relevant subgroups.

Consider Developing a Tier 1-D Predeployment Physical Fitness Assessment

Observations, discussions, and evaluations of deployment training suggest that the AF-FA may not be sufficient to ensure the physical fitness of airmen for advanced deployment to hostile or uncertain environments. Reports from training instructors suggest that up to 30 percent of airmen may not arrive to training with the requisite fitness to perform tasks effectively and with minimal risk of sustaining an injury. Despite these findings, there are several questions that need to be explored prior to implementation of a predeployment FA or Tier 1-D assessment. We have highlighted some of these questions in this section and provide recommended courses of action.

Who Should Be Required to Take a Tier 1-D Assessment?

Discussions with SMEs suggested that most deployment training is not physically demanding. Furthermore, a relatively small number of airmen attend advanced deployment training, which contains some very physically demanding tasks. Even though few airmen may attend these courses, some discussions with SMEs suggested that all airmen should be required to complete a Tier 1-D assessment. However, other SMEs noted that advanced deployment tasks (e.g., casualty drag) are low-probability events even for airmen deploying to hostile or uncertain environments. Given the resource requirements needed to successfully implement a new physical FA, we recommend limiting a Tier 1-D assessment to airmen who are likely to deploy to hostile or uncertain environments.

When and How Often Would This Assessment Be Administered?

Although it may be ideal to have all airmen physically fit and prepared prior to starting deployment training, some workshop SMEs suggested that a pretraining assessment may be difficult to implement given logistical challenges for scheduling and administration. As a first step, we recommend administering the assessment at the beginning and at the end of FC courses. This would not only provide a baseline to better understand fitness levels prior to training but also provide the data necessary to share with commanders on the physical fitness of airmen completing FC courses. The commanders with responsibility for these airmen could then use their FA scores to make informed decisions on whether the airmen should deploy. The AF should also continue to identify and implement different incentives (and consequences) to encourage airmen to maintain fitness levels required for deployment.

How Should the Tests and Standards Be Established?

Although we provide a preliminary analysis of advanced deployment training tasks, a more-thorough physical demands analysis is needed to guide a validation study that will establish the relationship between fitness tests and performance on training tasks. In short, we recommend that the AF follow processes that have been established for developing Tier 2 occupationally relevant fitness tests.

How Would a Tier 1-D Assessment Relate to the Current Air Force Fitness Assessment?

Considering the deployment cycles of airmen, we recommend an approach that incorporates a baseline standard and a predeployment standard (ready: operational deployment) as illustrated in Figure 4.1. The current AF-FA could serve as the baseline standard, and a Tier 1-D assessment could serve as the predeployment standard. However, according to AFI 36-2905 (2015), airmen require a current FA but do not need a passing score to be considered deployable. If the AF adopted a Tier 1-D assessment, all airmen considered deployable should, at a minimum, meet existing AF-FA standards, which would require a change in policy. Discussions with AF policy SMEs raised concerns that some airmen will attempt to game the system and fail the AF-FA on purpose to avoid a deployment. Consequently, we recommend evaluating various incentives (e.g., time off, extra pay) and consequences to ensure that airmen are sufficiently motivated to maintain the required fitness levels.

Figure 4.1. Example Physical Performance Levels Required Across an Air Force Career Life Cycle

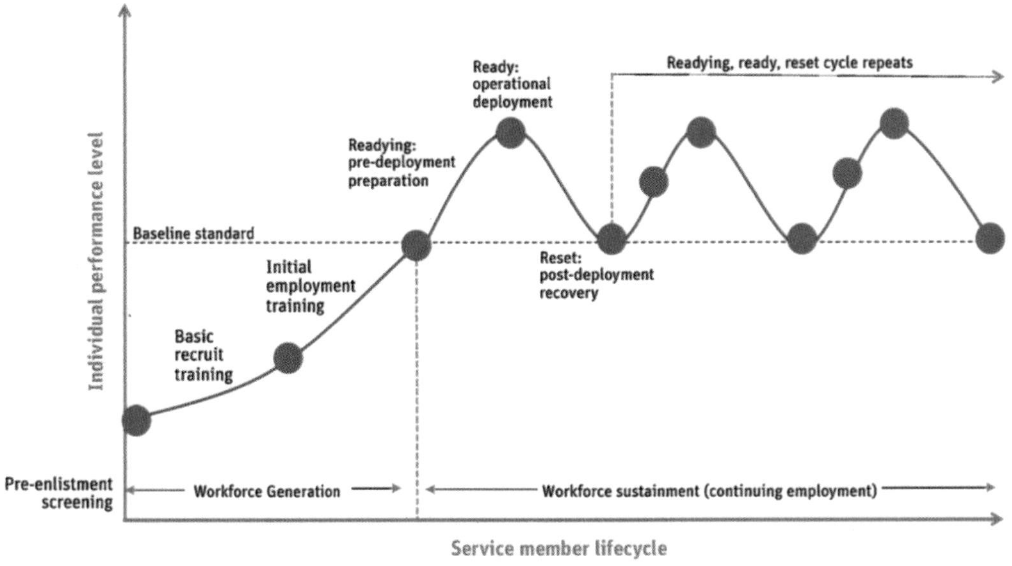

SOURCE: Billing and Drain, 2017.

Final Thoughts

This report summarizes published research on health-related fitness components and their relevance to the AF. The current AF-FA measures critical components of health-related fitness using well-supported assessments. However, the AF could further strengthen its FA by evaluating alternatives in a trial study. Additionally, the AF should consider developing a predeployment FA for airmen preparing to deploy to hostile or uncertain environments.

This report addresses important questions about whether the AF is using the right assessments. There are many other questions that the AF could explore to further strengthen the evidence for its AF-FA. For example, research is still needed to address whether assessment standards are appropriate for all airmen across relevant subgroups and whether those standards are being consistently enforced. Relatedly, research could also address which policies are most effective to encourage regular physical activity. Finally, there is a need to better understand the role of different levels of AF leadership (e.g., command, wing, squadron) on the physical fitness among today's airmen. For example, is the communication provided to airmen about the importance of fitness effective? Does the AF have a culture that supports and reinforces fitness? These are important issues that complement an AF-FA that is only administered to airmen once or twice a year.

Appendix A. Workshop Methodology

This appendix provides a brief overview of the methodology used for the workshops described in Chapter 3.

Participant Solicitation

To solicit participants, we worked with the ESU and the sponsor's office to identify relevant stakeholder organizations. Both workshops were held in June 2019. The Randolph Air Force Base workshop in Texas involved 11 participants. The participants represented a wide variety of perspectives, including exercise science, medical/injury prevention, health promotion, testing practice, training/services, National Guard, Reserves, data management, deployment, and safety. For the Arlington, Virginia, workshop at the Pentagon, five individuals participated. These participants represented fitness policy, exercise science, medical/injury prevention, AF senior enlisted leadership, and non-DoD exercise physiology.

Workshop Agenda

The project team developed a presentation and agenda for the workshop (see Table A.1 for the agenda). The primary purpose of the workshop was for experts on AF and military physical FA, policy, and requirements to review and discuss the relative strengths and weaknesses of (or gaps in) the AF-FA. The information they provided was used to identify strengths and limitations of different fitness tests within the AF-FA and to provide the broader policy context for applying FAs and standards in the AF.

Table A.1. Workshop Agenda

Time Frame	Activity	Facilitator and Participant
9:00–9:15 a.m.	Introductions	• RAND
9:15–10:15 a.m.	RAND and AF ESU briefings	• RAND • AF ESU
10:15–10:30 a.m.	Break	
10:30 a.m.–12:00 p.m.	Review and discussion of fitness components, assessments, and alternatives	• RAND • Discussion with workshop participants
12:00–1:00 p.m.	Break for lunch	
1:00–1:45 p.m.	Safety and injury prevention perspectives	• RAND • Discussion with workshop participants
1:45–2:30 p.m.	Guard and Reserve perspectives	• RAND • Discussion with workshop participants

Time Frame	Activity	Facilitator and Participant
2:30–2:45 p.m.	Break	
2:45–3:30 p.m.	Special topics	• RAND • Discussion with workshop participants
3:30–4:15 p.m.	Deployment perspectives	• RAND • Discussion with workshop participants
4:15–5:00 p.m.	Wrap-up	• RAND

NOTE: Workshop time frames are in eastern time for Arlington, Virginia, workshop and central time for Randolph workshop.

At the beginning of the workshop, the project team presented information on the project background and information from the team's literature review on the relative importance of different physical fitness components (e.g., CRE) to health, safety, and job performance outcomes. The team also provided examples of fitness tests (e.g., 1.5-mile run) used to assess each component and what evidence the literature provides on such factors as validity, reliability, perceived fairness, and ease of use. During the workshop, a senior-level team member facilitated the discussion while a junior team member typed notes to capture the discussion. These notes were later used by the team to identify themes from the workshop. See the following section for the protocol the team used to guide discussion and facilitate note-taking.

WORKSHOP PROTOCOL

[WELCOME, INTRODUCTIONS and BRIEF INTRO of RAND]

DESCRIPTION OF PROJECT

The Air Force has asked us to review the physical fitness assessment. As part of this review, we are asking subject-matter experts representing different perspectives (e.g., safety, medical, policy, exercise science) to provide inputs on the current fitness assessments and to evaluate potential alternatives that could be used in place of waivers, exemptions, and to address potential limitations associated with current assessments.

OBTAIN ORAL CONSENT

Before we begin, we need to get your consent to participate in this research.
[HANDOUT INFORMED CONSENT SHEETS] [READ ALOUD AND ASK IF THEY HAVE ANY QUESTIONS]

QUESTIONS

- What are the primary strengths of the current AF fitness assessment?
- What gaps or concerns do you have with the current AF fitness assessment?
    - PROMPTS: The current AF fitness assessment does not currently measure muscular strength. Is this a gap that needs to be addressed? If so, how?
- Waivers and exemptions are sometimes needed when environmental conditions prevent test administration or when an airman has an injury. In particular, some airmen receive waivers or exemptions for the 1.5-mile run.
    - What alternatives should the AF consider when environmental conditions prevent testing outdoors (e.g., snow, rain)?
    - ▪ PROMPTS: Shuttle Run, Step Test, Treadmill, Elliptical???
    - What alternatives should the AF consider when an airman is exempt due to an injury?
    - ▪ PROMPTS: Row Ergometer, Cycle Ergometer
- For each Air Force Fitness test component (1.5-mile run, sit-ups, push-ups, and the abdominal circumference), we will be asking you to consider the justification (e.g., scientific evidence) for inclusion, potential concerns or limitations, and feasibility of implementation (time to administer, cost, environment), and other factors to consider.

NOTE-TAKING FORM

Justification

Should the Air Force continue to use this test/assessment? What are the primary benefits of this assessment?

Concerns

What concerns or problems are associated with this test/assessment? Please consider risk of injury in taking the test or training for the test, fairness, reliability, etc.

Feasibility

What factors affect the feasibility of implementation of this test/assessment (time to administer, cost, environment)?

Alternative Assessments

What other assessments should the Air Force consider and why?

Please consider validity, reliability, feasibility of implementation (administration, cost), etc.

	Justification for Using	Concerns/Limitations	Feasibility
1.5-mile run			
Sit-ups			
Push-ups			
Abdominal Circumference			
Alternative Assessment #1			
Alternative Assessment #2			
Alternative Assessment #...			

CLOSING

Is there anything else you would like to add or do you have any questions about this project?

Thank you for your time.

Post-Workshop Review

The team used an iterative process to identify relevant themes from the workshop. Workshop notes were organized into the two workshops (Randolph Air Force Base in Texas and the Pentagon in Virginia). Two senior team members (one of whom facilitated the workshops) collectively reviewed workshop notes and identified key categories of topics recommended by participants to come up with an initial coding scheme. After creating the initial coding scheme, each of the two senior team members independently coded one workshop's notes by extracting text and applying primary and secondary (or "child") codes to each comment.[17] In some cases, a single comment may have been double- or triple-coded. After completing this initial coding exercise, each coder reviewed the other person's coding results and identified places where codes may have been unclear or misapplied. The two coders met to discuss and deconflict until they reached agreement. The final set of coded comments from the workshops was combined and used to discuss themes from the workshops.

It is important to note that in the main body of the report, we do not provide percentages or statistical estimates regarding the prevalence of workshop themes. Given the small number of workshops and the potential for mixed responses within a workshop, such percentages could be misleading. In addition, because not every participant responds to each question, it is not possible to provide precise statistical estimates regarding the percentage of individuals who make comments associated with a particular theme. Instead, the goal of these workshops was to gather detailed qualitative information from SMEs on strengths and limitations of the AF-FA.

[17] The coders did not use special coding software, such as NVivo or Dedoose, but instead used Microsoft Excel files to track codes. The coders determined that the small number of workshops and coders made the use of such software unnecessary.

Appendix B. Deployment Interviews

Deployment Task Interview Summary

Phone interviews with previously deployed airmen were conducted to identify physical tasks performed during recent deployments. Specifically, the questions focused on identifying and describing required tasks that were outside the duties required by their AFSC job specialty. Two hundred airmen received an email (see the following section) from the project sponsor requesting that they contact RAND to schedule an interview. Of those invited, 12 airmen responded and scheduled interviews with the data collection team.

Participants who contacted the research team were scheduled for 30-minute phone interviews and provided verbal consent before participating. The interview protocol, which includes the verbal consent information and rating scales, can be found in Appendix C. The first part of the interview was used to collect a variety of background and demographic information.

All 12 interviewees were active-duty AF personnel, with eight enlisted and ranking from E5 to E9 and four officers ranking from O3 to O6. Interviewees represented a variety of job specialties with no overlap among them. Nine were male, and three were female. Ten of the respondents had been deployed five or fewer times, while two had been deployed more than ten times each. All but three participants indicated that they had been deployed to a designated combat zone in the past. Seven participants indicated that their most recent deployment was in the Middle East or North Africa. Deployment locations indicated by the other five included Turkey, East Africa, West Africa, and Asia.

Participants were then asked to reflect on their most recent deployment and think about any physically demanding tasks performed that were not part of their career field duties. Participants were prompted to list all of the tasks they experienced that involved different movement patterns (e.g., lift, carry, push, pull).

In total, 25 separate tasks were described by participants, with seven participants describing two or more tasks, four describing one task, and one participant indicating that they did not perform any task outside their regular duties while deployed. While participants described each task, the interview team recorded which movement patterns were used. More than 50 percent of the tasks described involved some degree of lifting, carrying, bending, or walking, and 25 percent to 49 percent of the tasks included some degree of squatting, core rotation, or pulling (see Table B.1).

Table B.1. Movement Patterns Identified in Deployment Tasks

Movement Patterns	Percentage of Tasks
Lift; carry; bend; walk	50% or more
Squat; core rotation; pull	Between 25% and 49%
Climb; run; crawl; fast walk/jog	Less than 25%

Once the tasks had been described, participants were asked how important each task was to the mission, how often they performed each task, how much effort each task required to perform, and the amount of time required to complete. Participants were emailed documents containing the rating scales used in these questions for reference. Higher numeric ratings on all scales correspond to higher levels of that attribute. Importance, frequency, and duration ratings were based on a 5-point Likert-type scale, while intensity ratings were based on a 7-point Likert-type scale. The interview protocol in Appendix B contains the complete rating scales used during data collection. Each of the 25 tasks mentioned was grouped into one of five task categories. The most common tasks were gearing up and transport; moving equipment, materials, or supplies; and building, assembling, or performing maintenance. The tasks and their corresponding attribute ratings can be found in Table B.2.

Table B.2. Deployment Task Categories and Average Ratings

Deployment Task	*n*	Importance	Frequency	Duration	Intensity
Gearing up and transport	9	4.0	4.0	3.3	4.0
Moving equipment/materials/supplies	6	4.2	3.0	3.2	4.2
Building/assembly/maintenance	5	3.2	3.2	4.8	5.6
Evading enemy attack	3	4.3	3.0	1.3	5.0
Transporting wounded or deceased	2	5.0	2.5	1.5	4.0

In the final portion of the interview, participants were asked about specific factors encountered while working on these tasks that affected (increased or decreased) the level of effort required. By far the most common factor cited was heat, but other factors, such as terrain, cold, equipment or lack of equipment, weight of gear or materials, air pollution, and availability of help, were also mentioned more than once. At the conclusion of each interview, the interview team asked any needed clarification questions and thanked the respondents for their participation.

Email Invitation for Deployment Interview

From: Dunn, Troy (AF/A1P)
Cc: Robson, Sean
Subject: Evaluating the Air Force
Date: Monday, July 22, 2019

MEMORANDUM FOR AIRMEN WITH DEPLOYMENT EXPERIENCE FROM: AF/A1P
SUBJECT: Support for RAND

1. Request your support for an Air Force Director of Military Force Management Policy (AF/A1P) sponsored RAND study titled, "Evaluation of the Air Force Fitness Assessment."
2. In support of the National Defense Strategy, AF/A1P is sponsoring RAND research to evaluate how well the current Air Force Fitness Assessment supports and evaluates the health and readiness of airmen to meet the demands of current and future missions. Additional goals of this research are to identify potential alternate assessments for each component of the fitness assessment and develop policies that will promote physical readiness of airmen.
3. As part of this research effort, RAND will be conducting interviews with randomly selected airmen who have completed one or more deployments to better understand the tasks and physical demands associated with different mission environments. Because you have been identified as having met the interview criteria, we ask for your participation. These phone interviews are expected to last no more than 30 minutes.
4. Please contact RAND as soon as possible to schedule an interview.
    a. Mr. Matthew Strawn, RAND Corporation, Santa Monica, Calif., 90401; (310) 393-0411 ext. 6646; mstrawn@rand.org
    b. Dr. Sean Robson, Principal Investigator – RAND Corporation, Arlington, Va., 22202; (703) 413-1100 ext. 5528; smrobson@rand.org

Thank you in advance.
Brigadier General Troy E. Dunn

Appendix C. Interview Protocol

[WELCOME, INTRODUCTIONS and BRIEF INTRO of RAND]

Description of Project

The Air Force has asked us to review the physical fitness assessment. As part of this review, we are asking airmen who have recently deployed about their experiences performing physically demanding tasks that were not part of their regularly assigned AFSC duties. That is, tasks that were specific to a particular deployment.

Obtain Oral Consent

Before we begin, we need to get your consent to participate in this research.

[HANDOUT INFORMED CONSENT SHEETS] [READ ALOUD AND ASK IF THEY HAVE ANY QUESTIONS]

Oral Consent

Thank you for speaking with us. RAND is conducting a project sponsored by Brig Gen Troy Dunn, Director Force Management Policy (AF/A1P), to develop and recommend courses of action for potential improvements to the Air Force Fitness Assessment. As part of this project, we are holding discussions with airmen familiar with deployment training and deployment requirements. The goal of these discussions is to identify physical tasks and activities that airmen may be expected to perform during a deployment that are not part of their assigned occupational specialty.

This discussion should take about 30 minutes. Participation is voluntary. Please let us know now if you don't want to participate, or later if you want to stop participating at any time. You should feel free to skip any questions you prefer not to answer.

RAND will use the information from the discussion for project purposes only. Your identity will not be connected to your responses in any way. With the exception of the project team and anyone else who is part of today's discussion, your responses will not be known to others. We will be taking notes on the discussion, but to protect confidentiality, we will not include names or any other information that might identify you in our notes. We plan to use some comments/quotes from the discussion in reporting our findings and conclusions. However, all comments/quotes will be reported as anonymous and will not contain information that would lead you to be identified.

If you have any questions or comments about this RAND project, you can contact the project leader, Sean Robson (smrobson@rand.org; (703) 413-1100 x5528). If you have questions about your rights as a participant in this project or need to report a project-related injury or concern, you can contact RAND's Human Subjects Protection Committee toll-free at (866) 697-5620 or by emailing hspcinfo@rand.org. If possible, when you contact the Committee, please reference Study #2019-0096.

BACKGROUND QUESTIONS

1. Which component of the Air Force are you a part of?
    - Active-Duty Air Force
    - Air National Guard
    - Air Force Reserves
2. What is your current rank? ______________________________
3. What is your specialty? ______________________________
4. How many times have you been deployed? ____________
5. Have you previously been deployed to a designated combat zone?
    - Yes
    - No
6. What region was your most recent deployment?
    - Asia
    - Middle East/North Africa
    - Europe
    - Americas
    - Other
7. What is your gender?
    - Female
    - Male
    - Do not wish to answer

QUESTIONS

Please take a minute to think about a recent deployment.

- What types of tasks did you perform that were physically demanding that are not part of your career field duties?
- Describe any tasks you performed that required:

- Lifting?
- Carrying?
- Pushing or Pulling?
- Bending, Squatting or Kneeling?
- Walking?
- Running?
- Crawling?
- Climbing?
- Shoveling or Digging?
- Using Hand-Held Tools (powered or non-powered)?

TASK ANALYSIS

We now have a few follow-up questions about how important each task was to the mission, how often you performed the task, how much effort it required to perform, and the amount of time it took to complete. [READ BACK THE TASK TO MAKE SURE IT IS CLEAR AND THEN COLLECT RATINGS]

IMPORTANCE

How important is this task to effective mission performance?

1 = Not Important
2 = Somewhat Important
3 = Moderately Important
4 = Very Important
5 = Crucial

FREQUENCY

How often did you perform this task in your last deployment?

1 = Never Performed
2 = Seldom Performed (e.g., once or just a few times during deployment)
3 = Occasionally Performed (e.g., between once a week to a few times a month)
4 = Often Performed (e.g., at least a few times each week)
5 = Always Performed (e.g., daily to several times a day)

DURATION

In general, about how long does it take for you to perform this task before you can either rest or move on to a new task?

1 = 0 to 2 minutes
2 = Between 2 to 30 minutes
3 = Between 30 minutes to 1 hour
4 = Between 1 to 2 hours
5 = More than 2 hours

INTENSITY

How much physical effort is required to successfully complete this task? Physical effort is defined by strength, endurance, or movement quality (i.e., balance, flexibility).

1 Very, Very Light
2 Very Light
3 Light
4 Somewhat Hard
5 Hard
6 Very Hard
7 Very, Very Hard

AFTER TASK REVIEW: CONTRIBUTING FACTORS

As you considered these tasks, were there some specific factors that came to mind that increased or decreased the level of effort required? For instance, factors, such as heat, cold, light conditions, terrain, altitude; weight of gear carried; distance traveled; equipment—do they matter at all and if they do, to what extent? Are there other factors you can think of beyond these?

References

2018 Physical Activity Guidelines Advisory Committee, *2018 Physical Activity Guidelines Advisory Committee Scientific Report*, Washington, D.C.: U.S. Department of Health and Human Services, 2018. As of October 14, 2020: https://health.gov/sites/default/files/2019-09/PAG_Advisory_Committee_Report.pdf

ACSM—*See* American College of Sports Medicine.

AF—*See* U.S. Air Force.

AFI—*See* Air Force Instruction.

Air Force Instruction 10-405, *Expeditionary Readiness Training Program*, September 24, 2018.

Air Force Instruction 36-2905, *Fitness Program*, October 21, 2013, Incorporating Change 1, August 27, 2015.

Air Force Instruction 48-123, *Medical Examinations and Standards*, November 5, 2013.

Air Force Personnel Center, "Military Demographics," last updated January 1, 2020. As of October 19, 2020: https://www.afpc.af.mil/Portals/70/documents/03_ABOUT/Military%20Demographics%20Jan%202020.pdf?ver=2020-01-27-093137-550

American College of Sports Medicine, *ACSM's Health-Related Physical Fitness Assessment Manual*, 4th ed., Philadelphia, Pa.: Wolters Kluwer, 2013.

———, *ACSM's Health-Related Physical Fitness Assessment Manual*, 5th ed., Philadelphia, Pa.: Wolters Kluwer, 2017.

———, *ACSM's Guidelines for Exercise Testing and Prescription*, 10th ed., Philadelphia, Pa.: Wolters Kluwer, 2018.

Aragón, Luis F., "Evaluation of Four Vertical Jump Tests: Methodology, Reliability, Validity, and Accuracy," *Measurement in Physical Education and Exercise Science*, Vol. 4, No. 4, 2000, pp. 215–228.

Ashwell, Margaret, Philippa Gunn, and Sigrid Gibson, "Waist-to-Height Ratio Is a Better Screening Tool Than Waist Circumference and BMI for Adult Cardiometabolic Risk Factors: Systematic Review and Meta-Analysis," *Obesity Reviews*, Vol. 13, No. 3, 2011, pp. 275–286.

Aune, Dagfinn, Abhijit Sen, Manya Prasad, Teresa Norat, Imre Janszky, Serena Tonstad, Pål Romundstad, and Lars J. Vatten, "BMI and All Cause Mortality: Systematic Review and

Non-Linear Dose-Response Meta-Analysis of 230 Cohort Studies with 3.74 Million Deaths Among 30.3 Million Participants," *BMJ*, Vol. 353, No. 2156, May 4, 2016.

Axler, Craig T., and Stuart M. McGill, "Low Back Loads Over a Variety of Abdominal Exercises: Searching for the Safest Abdominal Challenge," *Medicine & Science in Sports & Exercise*, Vol. 29, No. 6, June 1997, pp. 804–811.

Barry, Vaughn W., Meghan Baruth, Michael W. Beets, J. Larry Durstine, Jihong Liu, and Steven N. Blair, "Fitness vs. Fatness on All-Cause Mortality: A Meta-Analysis," *Progress in Cardiovascular Diseases*, Vol. 56, No. 4, January–February 2014, pp. 382–390.

Bener, Abdulbari, Mohammad T. Yousafzai, Sarah Darwish, Abdulla O. A. A. Al-Hamaq, Eman A. Nasralla, and Mohammad Abdul-Ghani, "Obesity Index That Better Predict Metabolic Syndrome: Body Mass Index, Waist Circumference, Waist Hip Ratio, or Waist Height Ratio," *Journal of Obesity*, Vol. 2013, August 13, 2013.

Billing, Dan C., and Jace R. Drain, "International Congress on Soldiers' Physical Performance 2017: Research Priorities Across the Service Members Operational Lifecycle," *Journal of Science and Medicine in Sport*, Vol. 20, November 2017, pp. S1–S3.

Blair, Steven N., Harold W. Kohl III, Ralph S. Paffenbarger, Jr., Debra G. Clark, Kenneth H. Cooper, and Larry W. Gibbons, "Physical Fitness and All-Cause Mortality: A Prospective Study of Healthy Men and Women," *JAMA*, Vol. 262, No. 17, 1989, pp. 2395–2401.

Browning, Lucy M., Shiun Dong Hsieh, and Margaret Ashwell, "A Systematic Review of Waist-to-Height Ratio as a Screening Tool for the Prediction of Cardiovascular Disease and Diabetes: 0.5 Could Be a Suitable Global Boundary Value," *Nutrition Research Reviews*, Vol. 23, No. 2, December 2010, pp. 247–269.

Callander, Bruce D., "Jumper to Airmen: 'Get in Shape,'" *Air Force Magazine*, January 1, 2004.

Carlock, Jon M., Sarah L. Smith, Michael J. Hartman, Robert T. Morris, Dragomir A. Ciroslan, Kyle C. Pierce, Robert U. Newton, Everett A. Harman, William A. Sands, and Michael H. Stone, "The Relationship Between Vertical Jump Power Estimates and Weightlifting Ability: A Field-Test Approach," *Journal of Strength & Conditioning Research*, Vol. 18, No. 3, August 2004, pp. 534–539.

Caspersen, Carl J., Gregory M. Christenson, and Robert A. Pollard, "Status of the 1990 Physical Fitness and Exercise Objectives: Evidence from NHIS 1985," *Public Health Reports*, Vol. 101, No. 6, November–December 1986, pp. 587–592.

Caspersen, Carl J., Kenneth E. Powell, and Gregory M. Christenson, "Physical Activity, Exercise, and Physical Fitness: Definitions and Distinctions for Health-Related Research," *Public Health Reports*, Vol. 100, No. 2, March–April 1985, pp. 126–131.

CDC—*See* Centers for Disease Control and Prevention.

Centers for Disease Control and Prevention, “Defining Adult Overweight and Obesity,” webpage, last updated September 2020. As of October 14, 2020: https://www.cdc.gov/obesity/adult/defining.html

Childs, John D., Deydre S. Teyhen, Patrick R. Casey, Kimberly A. McCoy-Singh, Angela W. Feldtmann, Alison C. Wright, Jessica L. Dugan, Samuel S. Wu, and Steven Z. George, “Effects of Traditional Sit-Up Training Versus Core Stabilization Exercises on Short-Term Musculoskeletal Injuries in U.S. Army Soldiers: A Cluster Randomized Trial,” *Physical Therapy*, Vol. 90, No. 10, 2010, pp. 1404–1412.

Constable, Stefan, and Barbara Palmer, *The Process of Physical Fitness Standards Development*, Wright-Patterson Air Force Base, Ohio: Human Systems Information Analysis Center, 2000.

Cooper, Susan M., Julien S. Baker, Richard J. Tong, Elystan Roberts, and Michael Hanford, “The Repeatability and Criterion Related Validity of the 20 m Multistage Fitness Test as a Predictor of Maximal Oxygen Uptake in Active Young Men,” *British Journal of Sports Medicine*, Vol. 39, No. 4, April 1, 2005.

Courtright, Stephen H., Brian W. McCormick, Bennett E. Postlethwaite, Cody J. Reeves, and Michael K. Mount, “A Meta-Analysis of Sex Differences in Physical Ability: Revised Estimates and Strategies for Reducing Differences in Selection Contexts,” *Journal of Applied Psychology*, Vol. 98, No. 4, July 2013, pp. 623–641.

Crawford, Kim, Katelyn Fleishman, John P. Abt, Timothy C. Sell, Mita Lovalekar, Takashi Nagai, Jennifer Deluzio, Russell S. Rowe, Mark A. McGrail, and Scott M. Lephart, “Less Body Fat Improves Physical and Physiological Performance in Army Soldiers,” *Military Medicine*, Vol. 176, No. 1, January 2011, pp. 35–43.

Dalton, Mark, Adrian J. Cameron, Paul Z. Zimmet, Jonathan E. Shaw, Damien Jolley, David W. Dunstan, Timothy A. Welborn, and AusDiab Steering Committee, “Waist Circumference, Waist–Hip Ratio and Body Mass Index and Their Correlation with Cardiovascular Disease Risk Factors in Australian Adults,” *Journal of Internal Medicine*, Vol. 254, No. 6, December 2003, pp. 555–563.

de la Motte, Sarah J., Timothy C. Gribbin, Peter Lisman, Kaitlin Murphy, and Patricia A. Deuster, “Systematic Review of the Association Between Physical Fitness and Musculoskeletal Injury Risk: Part 2—Muscular Endurance and Muscular Strength,” *Journal of Strength and Conditioning Research*, Vol. 31, No. 11, November 2017, pp. 3218–3234.

Department of the Air Force, “ALO and TACP Tier 2 Operator Fitness Test Guidance Memorandum,” Washington, D.C., June 1, 2018.

Emerging Risk Factors Collaboration, “Separate and Combined Associations of Body-Mass Index and Abdominal Adiposity with Cardiovascular Disease: Collaborative Analysis of 58 Prospective Studies,” *Lancet*, Vol. 377, No. 9771, 2011, pp. 1085–1095.

Farrell, Steve, "Introduction to the Cooper Center Longitudinal Study (CCLS)," Cooper Institute, February 2, 2018. As of October 15, 2020: https://www.cooperinstitute.org/2018/02/02/introduction-to-the-cooper-center-longitudinal-study-ccls

Flegal, Katherine M., Brian K. Kit, Heather Orpana, and Barry I. Graubard, "Association of All-Cause Mortality with Overweight and Obesity Using Standard Body Mass Index Categories: A Systematic Review and Meta-Analysis," *JAMA*, Vol. 309, No. 1, 2013, pp. 71–82.

Friedl, Karl E., "Body Composition and Military Performance—Many Things to Many People," *Journal of Strength & Conditioning Research*, Vol. 26, No. 7, July 1, 2012, pp. S87–S100.

Friedl, Karl E., Joseph J. Knapik, Keijo Häkkinen, Neal Baumgartner, Herbert Groeller, Nigel A. S. Taylor, Antonio F. A. Duarte, Heikki Kyröläinen, Bruce H. Jones, William J. Kraemer, and Bradley C. Nindl, "Perspectives on Aerobic and Strength Influences on Military Physical Readiness: Report of an International Military Physiology Roundtable," *Journal of Strength & Conditioning Research*, Vol. 29, November 2015, pp. S10–S23.

Frith, Emily, and Paul D. Loprinzi, "Fitness Fatness Index and Alzheimer-Specific Mortality," *European Journal of Internal Medicine*, Vol. 42, July 2017a, pp. 51–53.

———, "The Protective Effects of a Novel Fitness-Fatness Index on All-Cause Mortality Among Adults with Cardiovascular Disease," *Clinical Cardiology*, Vol. 40, No. 7, July 2017b, pp. 469–473.

Gildea, Debbie, "Air Force Fitness Management System II Up, Accessible," Air Force Personnel Center Public Affairs, January 28, 2015. As of October 15, 2020: https://www.jbsa.mil/News/News/Article/598705/air-force-fitness-management-system-ii-up-accessible/

Global BMI Mortality Collaboration, Emanuele Di Angelantonio, Shilpa N. Bhupathiraju, David Wormser, Pei Gao, Stephen Kaptoge, Amy Berrington de Gonzalez, Benjamin J. Cairns, et al., "Body-Mass Index and All-Cause Mortality: Individual-Participant-Data Meta-Analysis of 239 Prospective Studies in Four Continents," *Lancet*, Vol. 388, No. 10046, August 20, 2016, pp. 776–786.

Granacher, Urs, Albert Gollhofer, Tibor Hortobágyi, Reto W. Kressig, and Thomas Muehlbauer, "The Importance of Trunk Muscle Strength for Balance, Functional Performance, and Fall Prevention in Seniors: A Systematic Review," *Sports Medicine*, Vol. 43, No. 7, 2013, pp. 627–641.

Griffith, J. R., Edward D. White III, R. David Fass, and Brandon M. Lucas, "Comparison of Body Composition Metrics for United States Air Force Airmen," *Military Medicine*, Vol. 183, No. 3–4, March–April 2018, pp. e201–e207.

Guh, Daphne P., Wei Zhang, Nick Bansback, Zubin Amarsi, C. Laird Birmingham, and Aslam H. Anis, "The Incidence of Co-Morbidities Related to Obesity and Overweight: A Systematic Review and Meta-Analysis," *BMC Public Health*, Vol. 9, No. 88, March 25, 2009.

Hardison, Chaitra M., Susan D. Hosek, and Chloe E. Bird, *Establishing Gender-Neutral Physical Standards for Ground Combat Occupations:* Volume 1. *A Review of Best-Practice Methods*, Santa Monica, Calif.: RAND Corporation, RR-1340/1-OSD, 2018. As of June 21, 2020: https://www.rand.org/pubs/research_reports/RR1340z1.html

Harvard Health Publishing, "Want a Stronger Core? Skip the Sit-Ups," webpage, last updated July 2, 2020. As of February 4, 2021: https://www.health.harvard.edu/staying-healthy/want-a-stronger-core-skip-the-sit-ups

Harwood, Georgina E., Mark P. Rayson, and Alan M. Nevill, "Fitness, Performance, and Risk of Injury in British Army Officer Cadets," *Military Medicine*, Vol. 164, No. 6, June 1999, pp. 428–434.

Hauret, Keith H., Ryan A. Steelman, Joseph R. Pierce, Joseph A. Alemany, Marilyn A. Sharp, Stephen A. Foulis, Jan E. Redmond, Maria C. Canino, Bruce S. Cohen, Michael S. McGurk, Whitfield B. East, and Bruce H. Jones, *Association of Performance on the Occupational Physical Assessment Test (OPAT), Injuries, and Attrition During Initial Entry Training—OPAT Phase 1*, Aberdeen Proving Ground, Md.: U.S. Army Public Health Center, PHR No. S.0047229-18b, January 10, 2018.

Hauschild, Veronique, David DeGroot, Shane Hall, Karen Deaver, Keith Hauret, Tyson Grier, and Bruce Jones, *Correlations Between Physical Fitness Tests and Performance of Military Tasks: A Systematic Review and Meta-Analyses*, Fort Belvoir, Va.: Defense Technical Information Center, 2014.

Hoffman, Jay R., Leah Chapnik, Ari Shamis, Uri Givon, and Benjamin Davidson, "The Effect of Leg Strength on the Incidence of Lower Extremity Overuse Injuries During Military Training," *Military Medicine*, Vol. 164, No. 2, February 1999, pp. 153–156.

Institute of Medicine, *Weight Management: State of the Science and Opportunities for Military Programs*, Washington, D.C.: National Academies Press, 2003.

———, *Fitness Measures and Health Outcomes in Youth*, Washington, D.C.: National Academies Press, 2012.

Janssen, Ian, Peter T. Katzmarzyk, and Robert Ross, "Body Mass Index, Waist Circumference, and Health Risk: Evidence in Support of Current National Institutes of Health Guidelines," *Archives of Internal Medicine*, Vol. 162, No. 18, 2002, pp. 2074–2079.

Jones, Bruce H., Matthew W. Bovee, John McA. Harris III, and David N. Cowan, "Intrinsic Risk Factors for Exercise-Related Injuries Among Male and Female Army Trainees," *American Journal of Sports Medicine*, Vol. 21, No. 5, September 1993, pp. 705–710.

Jorgensen, Torben, Lars B. Andersen, Karsten Froberg, Urs Maeder, Lisa von Huth Smith, and Mette Aadahl, "Position Statement: Testing Physical Condition in a Population—How Good Are the Methods?" *European Journal of Sport Science*, Vol. 9, No. 5, 2009, pp. 257–267.

Kaufman, Kenton R., Stephanie Brodine, and Richard Shaffer, "Military Training-Related Injuries: Surveillance, Research, and Prevention," *American Journal of Preventive Medicine*, Vol. 18, No. 3, April 2000, pp. 54–63.

Knapik, Joseph J., "The Importance of Physical Fitness for Injury Prevention: Part 2," *Journal of Special Operations Medicine: A Peer Reviewed Journal for SOF Medical Professionals*, Vol. 15, No. 2, Summer 2015, pp. 112–115.

Knapik, Joseph J., Shawn J. Scott, Marilyn A. Sharp, Keith G. Hauret, Salima Darakjy, William R. Rieger, Frank A. Palkoska, Stephen E. VanCamp, and Bruce H. Jones, "The Basis for Prescribed Ability Group Run Speeds and Distances in U.S. Army Basic Combat Training," *Military Medicine*, Vol. 171, No. 7, July 2006, pp. 669–677.

Knapik, Joseph J., Marilyn A. Sharp, Michelle Canham-Chervak, Keith G. Hauret, John F. Patton, and Bruce H. Jones, "Risk Factors for Training-Related Injuries Among Men and Women in Basic Combat Training," *Medicine and Science in Sports and Exercise*, Vol. 33, No. 6, June 2001, pp. 946–954.

Knudson, Duane, "Issues in Abdominal Fitness: Testing and Technique," *Journal of Physical Education, Recreation & Dance*, Vol. 70, No. 3, 1999, pp. 49–55.

Koch, Alexander J., Harold S. O'Bryant, Margaret E. Stone, Kim Sanborn, Christopher Proulx, Joe Hruby, Elizabeth Shannonhouse, Rhonda Boros, and Michael H. Stone, "Effect of Warm-Up on the Standing Broad Jump in Trained and Untrained Men and Women," *Journal of Strength & Conditioning Research*, Vol. 17, No. 4, November 2003, pp. 710–714.

Kodama, Satoru, Kazumi Saito, Shiro Tanaka, Miho Maki, Yoko Yachi, Mihoko Asumi, Ayumi Sugawara, Kumiko Totsuka, Hitoshi Shimano, Yasuo Ohashi, Nobuhiro Yamada, and Hirohito Sone, "Cardiorespiratory Fitness as a Quantitative Predictor of All-Cause Mortality and Cardiovascular Events in Healthy Men and Women: A Meta-Analysis," *JAMA*, Vol. 301, No. 19, May 20, 2009, pp. 2024–2035.

Leong, Darryl P., Koon K. Teo, Sumathy Rangarajan, Patricio Lopez-Jaramillo, Alvaro Avezum, Jr., Andres Orlandini, Pamela Seron, et al., "Prognostic Value of Grip Strength: Findings from the Prospective Urban Rural Epidemiology (PURE) Study," *Lancet*, Vol. 386, No. 9990, July 18, 2015, pp. 266–273.

Leu, John R., and Karl E. Friedl, "Body Fat Standards and Individual Physical Readiness in a Randomized Army Sample: Screening Weights, Methods of Fat Assessment, and Linkage to Physical Fitness," *Military Medicine*, Vol. 167, No. 12, December 2002, pp. 994–1000.

Lisman, Peter J., Sarah J. de la Motte, Timothy C. Gribbin, Dianna P. Jaffin, Kaitlin Murphy, and Patricia A. Deuster, "A Systematic Review of the Association Between Physical Fitness and Musculoskeletal Injury Risk: Part 1—Cardiorespiratory Endurance," *Journal of Strength & Conditioning Research*, Vol. 31, No. 6, June 2017, pp. 1744–1757.

Mahar, M. T., H. Sung, M. P. Whaley, S. Yun, W. J. Zhang, and D. N. Collier, "Relationships Among Fitness Measures and Health Outcomes in Youth," *Journal of Physical Activity & Health*, Vol. 22, Supp. 1, May 2014, pp. S171–S171.

Markovic, Goran, Drazan Dizdar, Igor Jukic, and Marco Cardinale, "Reliability and Factorial Validity of Squat and Countermovement Jump Tests," *Journal of Strength & Conditioning Research*, Vol. 18, No. 3, August 2004, pp. 551–555.

Martuscello, Jason M., James L. Nuzzo, Candi D. Ashley, Bill I. Campbell, John J. Orriola, and John M. Mayer, "Systematic Review of Core Muscle Activity During Physical Fitness Exercises," *Journal of Strength & Conditioning Research*, Vol. 27, No. 6, June 2013, pp. 1684–1698.

Mayorga-Vega, Daniel, Pablo Aguilar-Soto, and Jesús Viciana, "Criterion-Related Validity of the 20-M Shuttle Run Test for Estimating Cardiorespiratory Fitness: A Meta-Analysis," *Journal of Sports Science and Medicine*, Vol. 14, No. 3, September 2015, pp. 536–547.

Meadows, Sarah O., Charles C. Engel, Rebecca L. Collins, Robin L. Beckman, Matthew Cefalu, Jennifer Hawes-Dawson, Molly Waymouth, Amii M. Kress, Lisa Sontag-Padilla, Rajeev Ramchand, and Kayla M. Williams, *2015 Department of Defense Health Related Behaviors Survey (HRBS)*, Santa Monica, Calif.: RAND Corporation, RR-1695-OSD, 2018. As of December 12, 2020:
https://www.rand.org/pubs/research_reports/RR1695.html

McGill, Stuart M., "The Mechanics of Torso Flexion: Situps and Standing Dynamic Flexion Manoeuvres," *Clinical Biomechanics*, Vol. 10, No. 4, June 1995, pp. 184–192.

McGuire, Brian, Ross R. Vickers, Jr., John H. Reynolds, Anne Curry, Timothy Bockelman, and Ryan Massimo, *Examination of Pull-Ups and Push-Ups as Possible Alternatives to the Flexed Arm Hang on the Marine Corps Physical Fitness Test*, No. NHRC-11-21, San Diego, Calif.: Naval Health Research Center, 2011.

Moghaddam, Alireza Ansary, Mark Woodward, and Rachel Huxley, "Obesity and Risk of Colorectal Cancer: A Meta-Analysis of 31 Studies with 70,000 Events," *Cancer Epidemiology and Prevention Biomarkers*, Vol. 16, No. 12, December 2007, pp. 2533–2547.

Mongraw-Chaffin, Morgana L., Sanne A. E. Peters, Rachel R. Huxley, and Mark Woodward, "The Sex-Specific Association Between BMI and Coronary Heart Disease: A Systematic Review and Meta-Analysis of 95 Cohorts with 1.2 Million Participants," *Lancet Diabetes & Endocrinology*, Vol. 3, No. 6, June 2015, pp. 437–449.

Morrow, James R., Jr., Weimo Zhu, Don B. Franks, Marilu D. Meredith, and Christine Spain, "1958–2008: 50 Years of Youth Fitness Tests in the United States," *Research Quarterly for Exercise and Sport*, Vol. 80, No. 1, March 2009, pp. 1–11.

Myers, Jonathan, Amir Kaykha, Sheela George, Joshua Abella, Naima Zaheer, Scott Lear, Takuya Yamazaki, and Victor Froelicher, "Fitness Versus Physical Activity Patterns in Predicting Mortality in Men," *American Journal of Medicine*, Vol. 117, No. 12, 2004, pp. 912–918.

Nindl, Bradley C., "Strategies for Enhancing Military Physical Readiness in the 21st Century," Carlisle Barracks, Pa.: U.S. Army War College, 2012.

Nuzzo, James L., "The Case for Retiring Flexibility as a Major Component of Physical Fitness," *Sports Medicine*, Vol. 3, No. 10, December 16, 2019, pp. 853–870.

Nye, Nathaniel S., Mary T. Pawlak, Bryant J. Webber, Juste N. Tchandja, and Michelle R. Milner, "Description and Rate of Musculoskeletal Injuries in Air Force Basic Military Trainees, 2012–2014," *Journal of Athletic Training*, Vol. 51, No. 11, November 1, 2016, pp. 858–865.

Okorodudu, Dale O., Marwan Jumean, Victor M. Montori, Abel Romero-Corral, Virend K. Somers, Patricia J. Erwin, and Francisco Lopez-Jimenez, "Diagnostic Performance of Body Mass Index to Identify Obesity as Defined by Body Adiposity: A Systematic Review and Meta-Analysis," *International Journal of Obesity*, Vol. 34, No. 5, 2010, pp. 791–799.

Palmer, Barbara, Michael E. Rench, Jon W. Carroll, and Stefan H. Constable, *Health and Job-Specific Body Composition Standards for the US Air Force:* Volume 1, *Final Report*, Wright-Patterson Air Force Base, Ohio: Crew System Ergonomics Information Analysis Center, 2000.

Palmer, Barbara, and Jennifer Soest, *Expanded Air Force Physical Fitness Battery: Muscle Strength, Muscle Endurance, and Flexibility Considered*, Volume I, *Final Report*, Brooks Air Force Base, Tex.: Office for Prevention and Health Services Assessment, Armstrong Laboratory, October 1997.

Pate, Russell R., and Stephen Daniels, "Institute of Medicine Report on Fitness Measures and Health Outcomes in Youth," *JAMA Pediatrics*, Vol. 167, No. 3, March 2013, pp. 221–222.

Pate, Russell R., Maria Oria, and Pillsbury, Laura, eds., *Fitness Measures and Health Outcomes in Youth*, Washington, D.C.: National Academies Press, 2012.

Peterson, David D., "Proposed Performance Standards for the Plank for Inclusion Consideration into the Navy's Physical Readiness Test," *Strength & Conditioning Journal*, Vol. 35, No. 5, October 2013, pp. 22–26.

Prieske, Olaf, Thomas Muehlbauer, and Urs Granacher, "The Role of Trunk Muscle Strength for Physical Fitness and Athletic Performance in Trained Individuals: A Systematic Review and Meta-Analysis," *Sports Medicine*, Vol. 46, No. 3, March 1, 2016, pp. 401–419.

Ricciardi, Richard, Patricia A. Deuster, and Laura A. Talbot, "Effects of Gender and Body Adiposity on Physiological Responses to Physical Work While Wearing Body Armor," *Military Medicine*, Vol. 172, No. 7, July 2007, pp. 743–748.

Robson, Sean, Tracy C. Krueger, Jennifer L. Cerully, Stephanie Pezard, Laura Raaen, and Nahom M. Beyene, *Evaluating an Operator Physical Fitness Test Prototype for Tactical Air Control Party and Air Liaison Officers: A Preliminary Analysis of Test Implementation*, Santa Monica, Calif.: RAND Corporation, RR-2171-AF, 2018. As of June 22, 2020: https://www.rand.org/pubs/research_reports/RR2171.html

Robson, Sean, Maria C. Lytell, Carra S. Sims, Stephanie Pezard, Thomas Manacapilli, Amanda Anderson, Therese Bohusch, and Abigail Haddad, *Fit for Duty? Evaluating the Physical Fitness Requirements of Battlefield Airmen*, Santa Monica, Calif.: RAND Corporation, RR-618-AF, 2017. As of September 29, 2020: https://www.rand.org/pubs/research_reports/RR618.html

Rothman, Kenneth J., "BMI-Related Errors in the Measurement of Obesity," *International Journal of Obesity*, Vol. 32, No. 3, 2008, pp. S56–S59.

Roy, Tanja C., Thomas Songer, Feifei Ye, Ronald LaPorte, Tyson Grier, Morgan Anderson, and Michelle Chervak, "Physical Training Risk Factors for Musculoskeletal Injury in Female Soldiers," *Military Medicine*, Vol. 179, No. 12, December 2014, pp. 1432–1438.

Schmid, Daniela, and Michael F. Leitzmann, "Cardiorespiratory Fitness as Predictor of Cancer Mortality: A Systematic Review and Meta-Analysis," *Annals of Oncology*, Vol. 26, No. 2, February 2015, pp. 272–278.

Sloan, Robert A., Benjamin A. Haaland, Susumu S. Sawada, I-Min Lee, Xuemei Sui, Duck-Chul Lee, Yassine Ridouane, Falk Müller-Riemenschneider, and Steven N. Blair, "A Fit-Fat Index for Predicting Incident Diabetes in Apparently Healthy Men: A Prospective Cohort Study," *PloS One*, Vol. 11, No. 6, 2016.

Sloan, Robert A., Susumu S. Sawada, Corby K. Martin, and Benjamin Haaland, "Combined Association of Fitness and Central Adiposity with Health-Related Quality of Life in Healthy Men: A Cross-Sectional Study," *Health and Quality of Life Outcomes*, Vol. 13, No. 188, 2015.

Sloan, Robert A., Susumu S. Sawada, Lee I-Min, Yuko Gando, Ryoko Kawakami, Takashi Okamoto, Koji Tsukamoto, and Motohiko Miyachi, "The Association of Fit-Fat Index with Incident Diabetes in Japanese Men: A Prospective Cohort Study," *Scientific Reports*, Vol. 8, No. 569, January 2018.

Society for Industrial and Organizational Psychology, "Principles for the Validation and Use of Personnel Selection Procedures," *Industrial and Organizational Psychology*, Vol. 11, Supp. 1, December 2018, pp. 1–97.

Stevens, June, Jianwen Cai, Kelly R. Evenson, and Ratna Thomas, "Fitness and Fatness as Predictors of Mortality from All Causes and from Cardiovascular Disease in Men and Women in the Lipid Research Clinics Study," *American Journal of Epidemiology*, Vol. 156, No. 9, 2002, pp. 832–841.

Stodden, Dave, Ryan Sacko, and Danielle Nesbitt, "A Review of the Promotion of Fitness Measures and Health Outcomes in Youth," *American Journal of Lifestyle Medicine*, Vol. 11, No. 3, May–June 2017, pp. 232–242.

Strand, Sarah L., John Hjelm, Todd C. Shoepe, and Marie A. Fajardo, "Norms for an Isometric Muscle Endurance Test," *Journal of Human Kinetics*, Vol. 40, No. 1, 2014, pp. 93–102.

Suni, Jaana H., Pekka Oja, Raija T. Laukkanen, Seppo I. Mülunpalo, Matti E. Pasanen, Ilkka M. Vuori, Tuula-Marja Vartiainen, and Klaus Bös, "Health-Related Fitness Test Battery for Adults: Aspects of Reliability," *Archives of Physical Medicine and Rehabilitation*, Vol. 77, No. 4, 1996, pp. 399–405.

Tingelstad, Hans Christian, Daniel Theoret, Michael Spicovck, and Francois Haman, "Explaining Performance on Military Tasks in the Canadian Armed Forces: The Importance of Morphological and Physical Fitness Characteristics," *Military Medicine*, Vol. 181, No. 11–12, November–December 2016, pp. e1623–e1629.

U.S. Air Force Medical Support Agency, *Medical Standards Directory (MSD)*, Washington, D.C., September 10, 2018.

Vickers, Ross R., Jr., *Walk Tests as Indicators of Aerobic Capacity*, No. NHRC-02-22, San Diego, Calif.: Naval Health Research Center, 2002.

Vogel, James A., *A Review of Physical Fitness as it Pertains to the Military Services*, U.S. Army Research Institute of Environmental Medicine, Natick, Mass., July 1985.

Wagner, Dale R., and Vivian H. Heyward, "Techniques of Body Composition Assessment: A Review of Laboratory and Field Methods," *Research Quarterly for Exercise and Sport*, Vol. 70, No. 2, 1999, pp. 135–149.

Weiglein, Laura, Jeffery Herrick, Stacie Kirk, and Erik P. Kirk, "The 1-Mile Walk Test Is a Valid Predictor of VO_2 Max and Is a Reliable Alternative Fitness Test to the 1.5-Mile Run in U.S. Air Force Males," *Military Medicine*, Vol. 176, No. 6, June 1, 2011, pp. 669–673.

Wilkinson, David M., Sam D. Blacker, Victoria L. Richmond, Mark P. Rayson, and James L. J. Bilzon, "Relationship Between the 2.4-km Run and Multistage Shuttle Run Test Performance in Military Personnel," *Military Medicine*, Vol. 179, No. 2, February 2014, pp 203–207.

World Health Organization, "Health Promotion" webpage, undated. As of January 29, 2020: https://www.who.int/westernpacific/about/how-we-work/programmes/health-promotion

———, *Waist Circumference and Waist-Hip Ratio: Report of a WHO Expert Consultation, Geneva, 8–11 December 2008*, Geneva, Switzerland, 2011.

Yang, Justin, Costas A. Christophi, Andrea Farioli, Dorothee M. Baur, Steven Moffatt, Terrell W. Zollinger, and Stefanos N. Kales, "Association Between Push-Up Exercise Capacity and Future Cardiovascular Events Among Active Adult Men," *JAMA Network Open*, Vol. 2, No. 2, 2019.

Zaccardi, Francesco, Gary O'Donovan, David R. Webb, Thomas Yates, Sudhir Kurl, Kamlesh Khunti, Melanie J. Davies, and Jari A. Laukkanen, "Cardiorespiratory Fitness and Risk of Type 2 Diabetes Mellitus: A 23-Year Cohort Study and a Meta-Analysis of Prospective Studies," *Atherosclerosis*, Vol. 243, No. 1, November 2015, pp. 131–137.